Elegant Fractals

Automated Generation of Computer Art

Fractals and Dynamics in Mathematics, Science, and the Arts: Theory and Applications

ISSN: 2382-6320

Published

Vol. 2   *Elegant Fractals: Automated Generation of Computer Art*
by Julien Clinton Sprott (University of Wisconsin-Madison, USA)

Vol. 1   *Benoit Mandelbrot: A Life in Many Dimensions*
edited by Michael Frame (Yale), Nathan Cohen (Fractal Antenna Systems, Inc., USA)

Forthcoming

Vol. 3   *Fractalize That: A Visual Essay on Statistical Geometry*
by John Shier

Vol. 4   *Quantized Number Theory, Fractal Strings and the Riemann Hypothesis: From Spectral Operators to Phase Transitions and Universality*
by Hafedh Herichi (Santa Monica College, USA), Michel L Lapidus (University of California, Riverside, USA)

Fractals and Dynamics in Mathematics, Science, and the Arts: Theory and Applications

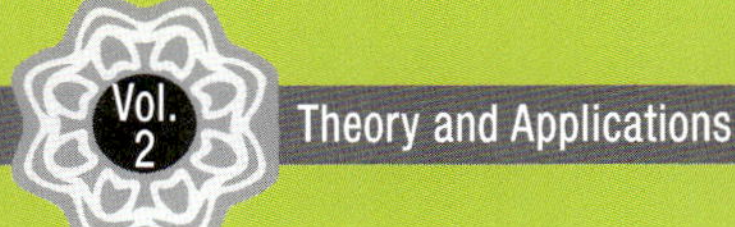

Elegant Fractals

Automated Generation of Computer Art

Julien Clinton Sprott

University of Wisconsin–Madison, USA

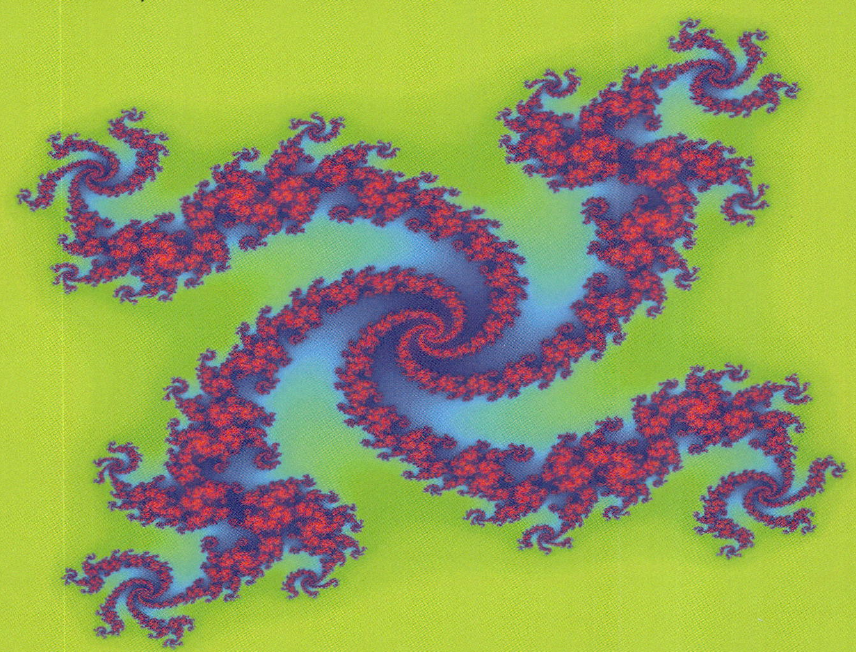

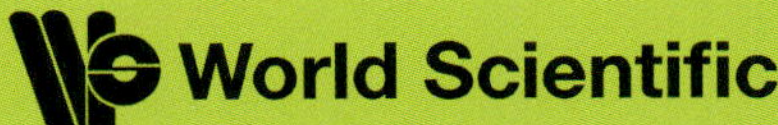

World Scientific

NEW JERSEY · LONDON · SINGAPORE · BEIJING · SHANGHAI · HONG KONG · TAIPEI · CHENNAI · TOKYO

Published by

World Scientific Publishing Co. Pte. Ltd.

5 Toh Tuck Link, Singapore 596224

USA office: 27 Warren Street, Suite 401-402, Hackensack, NJ 07601

UK office: 57 Shelton Street, Covent Garden, London WC2H 9HE

British Library Cataloguing-in-Publication Data
A catalogue record for this book is available from the British Library.

**Fractals and Dynamics in Mathematics, Science, and the Arts:
Theory and Applications — Vol. 2
ELEGANT FRACTALS
Automated Generation of Computer Art**

Copyright © 2019 by World Scientific Publishing Co. Pte. Ltd.

ISBN 978-981-3237-13-1

For any available supplementary material, please visit
https://www.worldscientific.com/worldscibooks/10.1142/10906#t=suppl

Printed in Singapore

Dedicated to the memory of
Benoit Mandelbrot

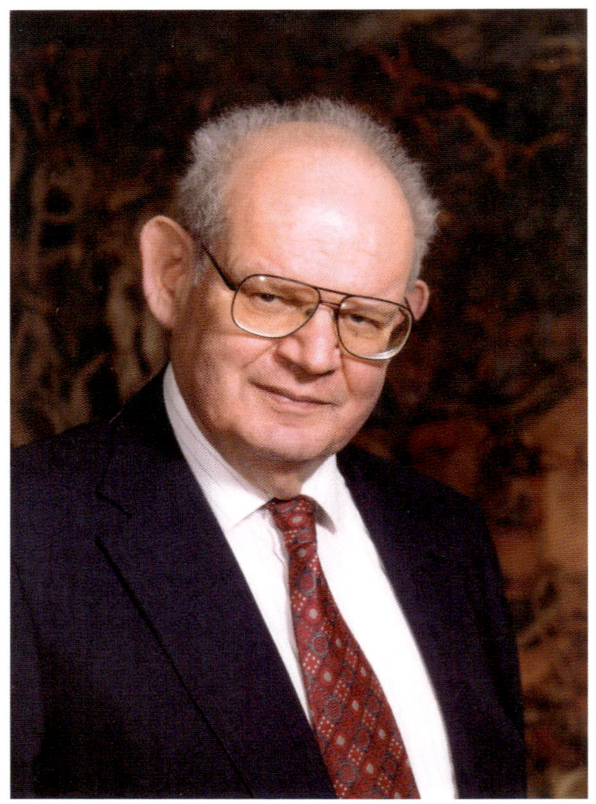

Benoit Mandelbrot, 1924–2010

Photo: http://users.math.yale.edu/mandelbrot/photos.html

Preface

Although fractals have been known and studied by many people for over a century, their current interest is due largely to Benoit Mandelbrot who coined the term 'fractal' in 1975 [Albers and Alexanderson (2008)] and whose book *The Fractal Geometry of Nature* [Mandelbrot (1982)] brought them into the mainstream of popular and professional mathematics. His role as 'the father of fractals' parallels and complements the role that Edward Lorenz played as 'the father of chaos.' In fact, some of the most important examples of fractals come from the chaotic solutions of equations and the natural processes that they model. Ruth Richards calls fractals 'the fingerprints of chaos' [Richards (1999)], although they can be produced in many other ways. The advent of modern computers has made it possible to produce and visualize fractals for both scientific and artistic purposes.

From the start of my interest in chaos and fractals thirty years ago, I have been intrigued by the enormous variety of fractal patterns that can be produced by simple equations. I wrote one of the first computer programs to search through the parameter space of those equations and identified thousands of examples of fractal strange attractors never before seen or studied [Sprott (1993a,b)]. Over the years, as computers grew ever more powerful and capable of producing high-resolution color graphics, I continually improved my programs for finding chaos in systems where it had not been previously known and in making increasingly elegant fractal art.

Ten years ago, I began writing the companion book, *Elegant Chaos: Algebraically Simple Chaotic Flows* [Sprott (2010)], after which this book is patterned. In *Elegant Chaos* the elegance is in the form of the equations that produce the attractors and is precisely defined. Here, the elegance refers to the aesthetics of the fractal pattern and is purely subjective. Nevertheless, the search methods are similar, with the only difference being in what is optimized. This book is an attempt to share some of those methods so that others can exploit them and to show new examples of the elegant fractals that lurk in the solutions of simple equations.

This book should be of interest to researchers looking for mathematical systems with 'interesting' attractors or to hobbyists or artists looking for a method to generate an endless variety of artistic patterns. It should also be useful to students and their instructors as an engaging way to learn about chaotic dynamics and fractal geometry as well as computer programming. The book assumes only an elementary knowledge of calculus. The systems are iterated maps and initial-value ordinary differential equations, as well as some examples of cellular automata.

You will get the most out of this book if you can write simple computer programs in the language of your choice or have access to software that allows you to solve systems of coupled equations and to display the results graphically. All the calculations and figures in this book were done with the *PowerBASIC Console Compiler* (http://www.powerbasic.com/), and are freely available at http://sprott.physics.wisc.edu/edu/fractals/elegant/. There is no substitute for the thrill and insight of seeing the solution of a simple equation unfold as the trajectory wanders in real time across your computer screen using a program of your own making. A goal of this book is to inspire and delight as well as to teach. I hope you will enjoy reading and studying it as much as I did writing it.

Many people have contributed to the ideas in this book. My greatest debt is to George Rowlands who lured me away from plasma physics when he got me interested in chaos and fractals in the late 1980s. I am also grateful to Ted Pope who assured me that these patterns have artistic merit and to Cliff Pickover who has inspired and encouraged me for many years with his own incredible computer graphics. My colleague and friend, Robin Chapman, has provided suggestions and encouragement including coauthoring *Images of a Complex World: The Art and Poetry of Chaos* [Chapman and Sprott (2005)] which exploits these ideas but is aimed at a broader and less technical audience. Bill Hoover read a draft of the manuscript and offered helpful suggestions. Finally, I thank the many colleagues and students with whom I have coauthored papers about chaotic systems that were found and optimized using my computer programs.

J. C. Sprott
Madison, Wisconsin
June, 2018

Contents

Chapter 1

Preliminaries

This chapter will introduce some of the fractal and chaos concepts and terminology so that the book is self-contained and readable even if you are unfamiliar with the subject. It will also describe how the images are produced using the Hénon map as the main example. It should enable you to get a quick start producing your own elegant fractal images.

1.1   Fractals

Fractals have existed since the beginning of time and are all around us, but only in the last few decades have they been widely recognized and properly described. A fractal is a geometric object that contains miniature copies of itself and is thus said to be *self-similar*. In fact, a true mathematical fractal contains infinitely many copies of itself on ever smaller scales. Mathematical fractals, just like mathematical points and lines, have never been seen and exist only in the minds of mathematicians. When you draw a point or a line, you give it a small width to make it visible, and in so doing it is no longer a true point or line. So it is with the fractal objects in nature and the images that constitute the subject of this book.

A whimsical example of the kind of fractal we will be considering is the word 'FRACTAL' in Fig. 1.1, which is only visible because the process of miniaturization is stopped before the infinitely many points that make up the image become too small to see. Technically, it is called a *prefractal*, and all fractals in nature are of this sort. Even worse, most natural fractals are self-similar only in a rough statistical sense. You will never find exact copies of the whole upon magnifying a portion of them. Clouds, mountains, and rivers are common examples of fractals in nature.

Other examples of natural objects that are usually called fractals are

1

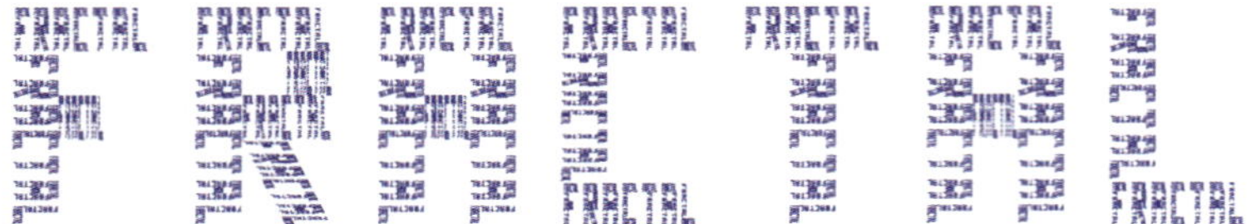

Fig. 1.1   A self-similar geometric object (a fractal).

trees and coastlines, shown in stylistic form in Figs. 1.2 and 1.3, respectively. In a tree, the branches have branches which have branches, *ad infinitum.* A coastline or a river has bends that have smaller bends, *ad infinitum.* A coastline is not a line at all, and any attempt to measure its length will frustrate you because the result will depend on the length of the ruler you use to measure it, and your estimate of the length will grow ever larger without limit as you improve your measurement. The object in Fig. 1.3 is called a *Koch snowflake* or *Koch island*, named after the Swedish mathematician Helge von Koch (1904), and we will return to it toward the end of Chapter 9.

Another classic example of a mathematical fractal is the *Sierpiński carpet* (or *Sierpiński gasket*) shown in Fig. 1.4, which was first described by the Polish mathematician Wacław Sierpiński (1916). It is not a very good carpet to cover your floor because it has infinitely many holes of arbitrarily small size. In fact, there is a sense to be described shortly in which it is all holes. Every point on the carpet is arbitrarily close to a hole, and the carpet itself has zero area. If you were to have such a carpet made for your bedroom, the material cost should be zero, but the labor cost would be infinite!

The previous examples of fractals are embedded in the two-dimensional plane. Fractals can also be embedded in higher dimensions. Indeed, a fractal tree usually lives in three-dimensional space, and so the image in Fig. 1.2 might better be considered as the shadow of a tree on the ground.

The *Menger sponge* [Menger (1928)] in Fig. 1.5 is an example similar to the Sierpiński carpet but embedded in three-dimensions. In fact, each face of the sponge is a Sierpiński carpet. The Menger sponge has infinite surface area but zero volume. Fractals can also be embedded in dimensions higher than three, and we will soon discuss ways to display and visualize them.

Objects such as these were proposed and studied over a hundred years ago, but even most mathematicians considered them to be monstrosities of no practical interest. Only in the 1980s did Mandelbrot (1982) bring

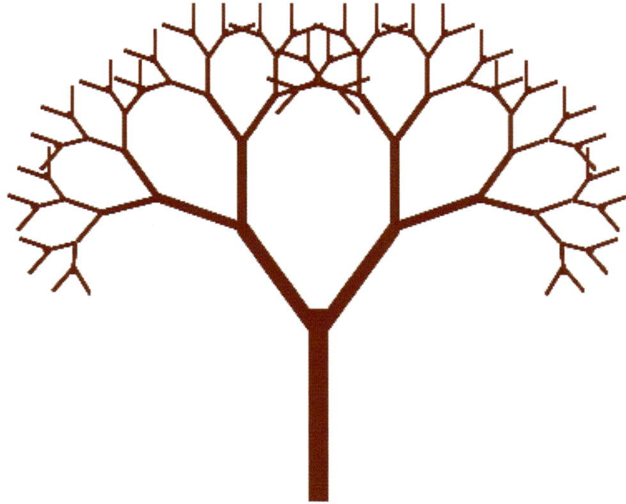

Fig. 1.2   A stylized fractal tree.

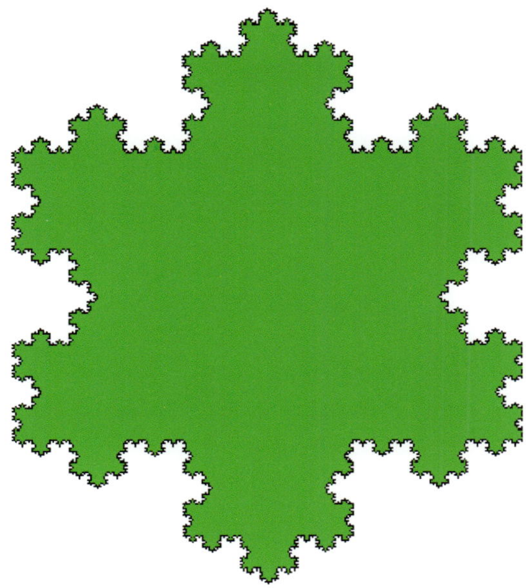

Fig. 1.3   A stylized fractal coastline.

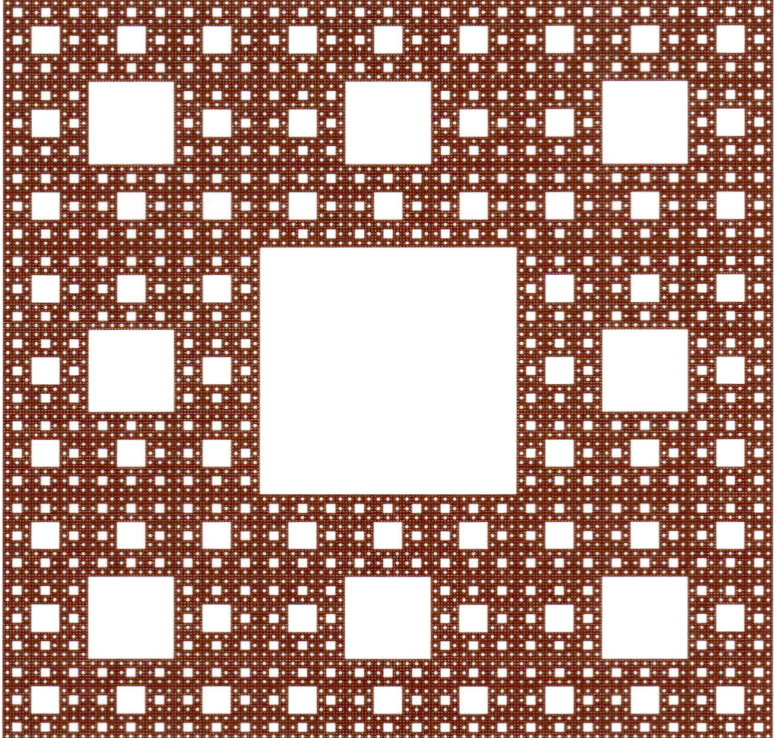

Fig. 1.4   Sierpiński carpet (a classic fractal object).

them to the attention of a wide audience and demonstrate their relevance
to patterns and processes that are ubiquitous in nature.

1.2   Two-dimensional Maps

Whatever images you produce will presumably need to be displayed on a
two-dimensional computer screen or piece of paper, although 3-D printers
raise additional interesting possibilities. Thus it is reasonable to begin by
considering a two-dimensional *iterated map*,

$$x_{n+1} = f(x_n, y_n)$$
$$y_{n+1} = g(x_n, y_n),$$

(1.1)

where x will be taken as the horizontal position of a point in the image and
y will be taken as the vertical position of that point. Beginning with some

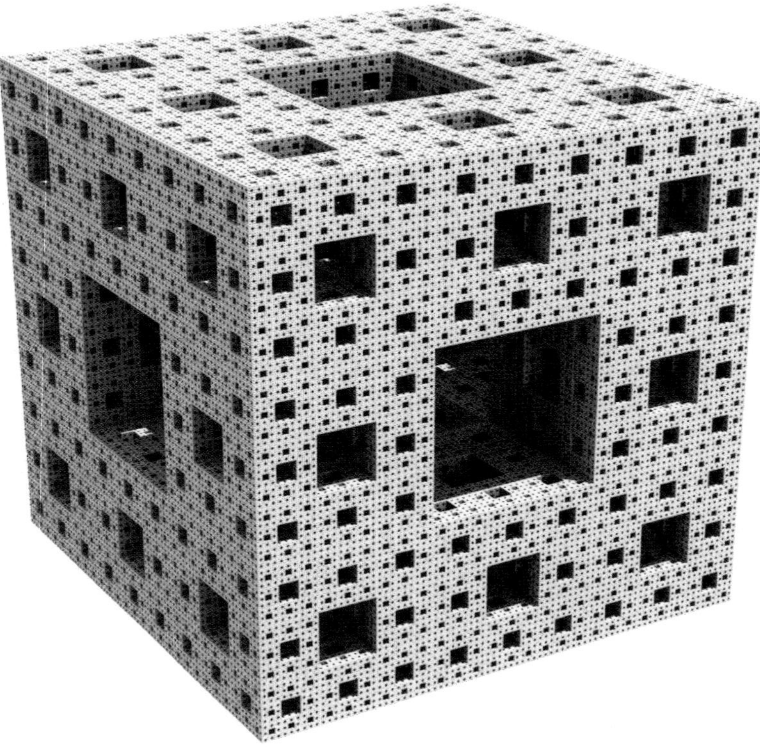

Fig. 1.5   Menger sponge (a classic fractal object embedded in three dimensions).

initial condition (x_0, y_0), the equations are repeatedly solved (*iterated*) to determine the position of successive points (x_n, y_n) in a kind of mathematical feedback operation. The functions f and g determine the shape of the resulting object after plotting the points for several million such iterations. Interesting patterns are produced only when at least one of the functions is *nonlinear*, and even then, it is far from guaranteed.

If f and g are *linear functions* of the form $f, g = ax_n + by_n + c$, successive iterates will either approach a point in the plane (a *stable fixed point*) or will wander off to infinity, neither of which will produce a fractal pattern. The nonlinearity is necessary to allow x and y to grow but to make the orbit fold back on itself rather than going to infinity. We say that such an orbit is *bounded*.

1.3   The Hénon Map

Consider the famous example of the Hénon map [Hénon (1976)] in which $f(x,y) = 1 - 1.4x^2 + 0.3y$ and $g(x,y) = x$ giving

$$x_{n+1} = 1 - 1.4x_n^2 + 0.3y_n$$
$$y_{n+1} = x_n, \tag{1.2}$$

which can be written in a more compact form as

$$x_{n+1} = 1 - 1.4x_n^2 + 0.3x_{n-1}. \tag{1.3}$$

Systems that can be written in such a form are called *scalar time-delay maps* since they involve a single scalar variable x and its values at some number of previous times. A *scalar* is a quantity that is described by a single number as opposed to a *vector* that requires more than one such number. For example, temperature is a scalar, but wind velocity is a vector since it has both a magnitude and direction.

In the Hénon map, the single nonlinearity is in the x^2 term, and thus it is the algebraically simplest two-dimensional map whose iterates produce a fractal pattern, a so-called *quadratic map*. For most of the examples in this book, any small values will suffice as initial conditions, and unless otherwise noted, they will be taken as 0.1 since zero will often be an *equilibrium point* from which the orbit cannot escape, although that is not the case for the Hénon map. For Eq. (1.2) with $x_0 = 0.1$ and $y_0 = 0.1$, the next few iterates are given by

$$
\begin{aligned}
x_1 &= 1.016 & y_1 &= 0.1 \\
x_2 &= -0.4151584 & y_2 &= 1.016 \\
x_3 &= 1.06350090407322 & y_3 &= -0.4151584 \\
x_4 &= -0.707995362150367 & y_4 &= 1.06350090407322 \\
x_5 &= 0.617289865264964 & y_5 &= -0.707995362150367 \\
x_6 &= 0.254135902492518 & y_6 &= 0.617289865264964 \\
x_7 &= 1.09476787986953 & y_7 &= 0.254135902492518 \\
x_8 &= -0.601682624363875 & y_8 &= 1.09476787986953 \\
x_9 &= 0.821599591314898 & y_9 &= -0.601682624363875 \\
x_{10} &= -0.125541031137494 & y_{10} &= 0.821599591314898
\end{aligned}
$$

The iterates continue to fluctuate between the limiting values of $x_{min} = -1.2846638$ and $x_{max} = 1.27297362$ without ever repeating for as long as you care to iterate. Note that y is just the previous value of x, and so they have the same limiting values ($y_{min} = x_{min}$ and $y_{max} = x_{max}$). It

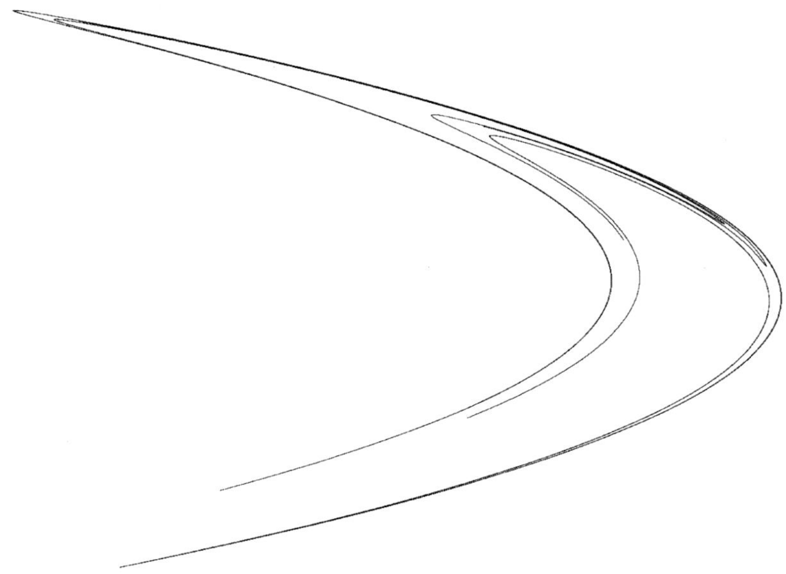

Fig. 1.6   Strange attractor for the Hénon map from Eq. (1.2).

is useful to determine the limiting values (x_{min} and x_{max}) before plotting the solution so that the plot fills the screen. If the screen has a width w and height h (in pixels), a point at (x, y) is plotted at the pixel whose coordinates (x_p, y_p) are given by

$$x_p = w(x - x_{min})/(x_{max} - x_{min})$$
$$y_p = h(y_{max} - y)/(y_{max} - y_{min}). \tag{1.4}$$

The sign reversal in y_p assumes that the pixel value increases in the downward direction as in common in computer languages, which is a convention that dates back to the days of teletype machines where the output was printed on a long continuous roll of paper.

Figure 1.6 shows the first ten million iterates of Eq. (1.2) plotted in this way. The resulting object is a type of fractal called a *strange attractor*. Technically, this is only a portion of the strange attractor, since infinitely many iterations would be required to cover it completely with points of infinitesimal size. Since the points are plotted with a finite size, the two are indistinguishable to the resolution of the plot, and thus we will refer to such images as *attractors*.

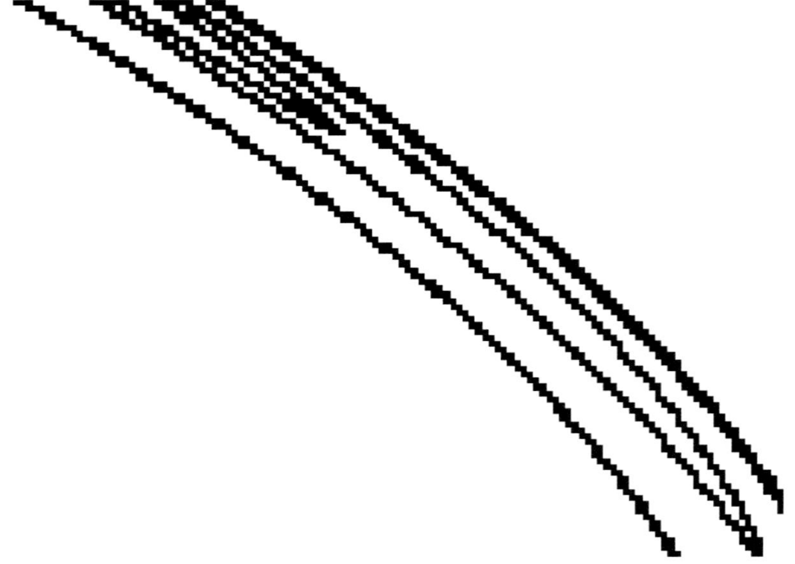

Fig. 1.7　A portion of the Hénon attractor enlarged by a factor of ten without increasing the resolution showing ragged lines that are a plotting artifact.

1.4　Anti-aliasing

To examine the fractal nature of the Hénon map, it is useful to enlarge a portion of its attractor on the upper right edge by a factor of ten in each direction, giving the plot in Fig. 1.7. You will notice immediately the ragged nature of the lines which is entirely a plotting artifact. The reason is that the image in Fig. 1.6 was made at a resolution of 300 dots per inch (1 inch = 2.54 cm), which corresponds to 1200 horizontal pixels by 900 vertical pixels to fit on the computer screen. The corresponding enlargement in Fig. 1.7 is thus at a resolution of 30 dots per inch (120 × 90 pixels). Of course you can simply increase the resolution of these plots, and sometimes that is reasonable and desirable, but it will increase the computation time and will make some of the lines too thin to see. Fortunately, there is a better way, called *anti-aliasing* [Freeman (1974)] that we will discuss before proceeding.

　　The problem is that each pixel in Fig. 1.7 is a square 1/30 of an inch on a side, and the entire pixel is either black or white. Thus the individual iterates must be converted to a pair of integers, and the entire pixels corresponding to those integers are set to black. This is fortunate because if you plotted the iterates as mathematical points, they would be invisibly small, and even with infinitely many of them, they would still be invisible.

However, the pixels on a computer screen typically consist of three colors (red, green, and blue), each of which can take on integer intensity values of 0 to 255 (one *byte* for each). When all three have a value of 255, the pixel is white, and when all three have a value of 0, the pixel is black. Thus you have 256 levels of gray to exploit with equal intensity for the three primary colors. In the interest of conserving ink, we will plot most of the images on a white background, although for some artistic purposes, you may want to choose a different background color, put a border around the image, include your name and date, show the equation that produced it, or add other embellishments.

The idea behind anti-aliasing is to make a pixel completely black if the point to be plotted falls at its center, but if the point is off center, the pixel and three of its nearest neighbors are plotted in a shade of gray that depends on how far off center the point is. The result is to smooth the jaggedness of the lines but at the expense of blurring them slightly.

Suppose you want to plot a point at pixel position (x_p, y_p), where x_p and y_p are not integers. You would take the integer part of each (the largest integers that are smaller than x_p and y_p, often called the *floor*), denoted by $\lfloor x_p \rfloor$ and $\lfloor y_p \rfloor$, and the fractional parts (the *remainders*), denoted by $\delta x_p = x_p - \lfloor x_p \rfloor$ and $\delta y_p = y_p - \lfloor y_p \rfloor$, and assign the following gray levels to the four pixels nearest to each point:

Pixel	Gray level
$(\lfloor x_p \rfloor, \lfloor y_p \rfloor)$	$255 - 255\delta x_p \delta y_p$
$(\lfloor x_p \rfloor, \lfloor y_p \rfloor + 1)$	$255 - 255\delta x_p(1 - \delta y_p)$
$(\lfloor x_p \rfloor + 1, \lfloor y_p \rfloor)$	$255 - 255(1 - \delta x_p)\delta y_p$
$(\lfloor x_p \rfloor + 1, \lfloor y_p \rfloor + 1)$	$255 - 255(1 - \delta x_p)(1 - \delta y_p).$

However, you should not plot a pixel with a lighter shade of gray (a larger gray level) than it already has. If you want to plot the point in color, just apply the method to each of the three color components separately, replacing the 255 with the value for that component of the color.

The anti-aliasing method is illustrated in Fig. 1.8 where the pixels are enlarged to make them easily visible. For example, to plot a point at a pixel position given by $(x_p, y_p) = (1.2, 1.4)$, the usual method would be simply to round the values to $(1, 1)$ and plot that pixel alone. Instead, you can construct a square centered on the point with a size equal to that of a

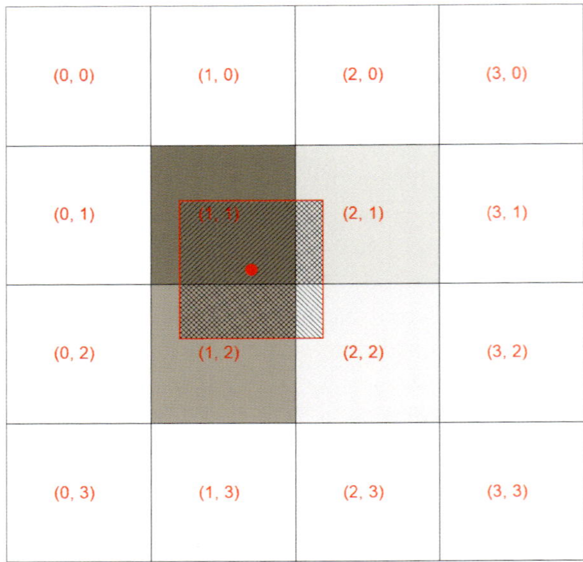

Fig. 1.8   With anti-aliasing, a point such as $(1.2, 1.4)$ that does not fall at the center of a pixel is plotted using the four nearest pixels but with a gray scale determined by its proximity to the center of each pixel.

single pixel as indicated in red. Then use the fraction of the area of that square that overlaps each pixel to determine the corresponding gray level according to the above rules.

Figure 1.9 shows the same portion of the attractor as Fig. 1.7 but plotted with anti-aliasing as described above. The edges are smoother, but the points are now spread over twice as many pixels in each direction with some blurring. You can reduce the spreading by increasing the resolution as shown in Fig. 1.10 which has 300 dots per inch. When combined with anti-aliasing, this resolution turns out to be a good compromise between having ragged edges at lower resolution and having parts of the image fade into invisibility at higher resolution. It also allows you to view the image on most computer screens without shrinking its size or scrolling. Therefore, most of the remaining images in this book will be plotted in this way. It is a simple matter to increase the resolution as desired for other purposes such as making large prints or posters, and some such examples are shown at the end of each chapter.

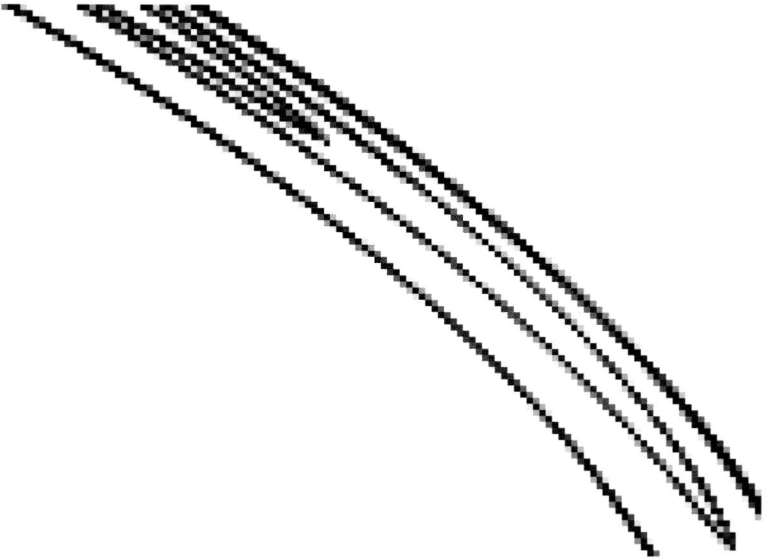

Fig. 1.9   The same plot as in Fig. 1.7 but with anti-aliasing.

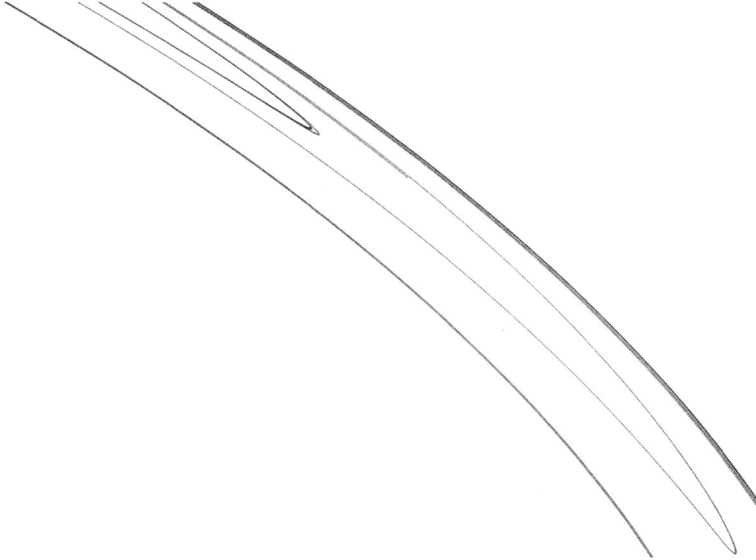

Fig. 1.10   The same plot as in Fig. 1.9 but with ten times the resolution, showing the fractal structure of the Hénon attractor.

1.5   Strange Attractors

The Hénon map in Fig. 1.6 is an example of a *strange attractor*. It is called 'strange' because it is a *fractal*. It is neither a very long line wrapped back on itself many times, nor is it a surface of finite area with many holes throughout. Rather it is an object that contains infinitely many copies of itself on ever smaller scales, the first few of which are evident in Fig. 1.10 which is an enlargement of a portion of Fig. 1.6 by a factor of ten in each direction.

Further enlargement of Fig. 1.10 would reveal that the wide line at the right of the group of three lines at the right edge actually consists of three closely spaced lines, the rightmost of which consists of three even more closely spaced lines, and so forth to whatever magnification you employ. Furthermore, the middle of those three lines is actually two lines, each of which is two lines, *ad infinitum.*

Since the object is neither a long line nor a surface with many holes, its *fractal dimension* is neither 1 nor 2, but rather it has an intermediate value of 1.2583 according to one method of calculation (the *Kaplan–Yorke dimension* [Kaplan and Yorke (1979)]). Just as the area of a mathematical line is zero because of its infinitesimal width, the area of the Hénon attractor is zero even after infinitely many iterations. That means if you were to step back and throw perfectly sharp darts at it, you would never hit it, but your darts would always strike one of the infinitely many holes.

This is the same reason it would be invisible, just as would be a line, if plotted at very high resolution. In fact, any object with a fractional dimension less than 2.0 will have zero area and will consist 100% of holes and thus would be invisible if plotted at high resolution. The Sierpiński carpet with a dimension of $\log(8)/\log(3) \approx 1.8928$ in Fig. 1.4 and the Menger sponge with a dimension of $\log(20/\log(3) \approx 2.7268$ in Fig. 1.5 are other such examples.

The object is an *attractor* in the sense that there is an infinite set of points representing possible initial conditions, all of which asymptotically approach the attractor upon repeated iteration of the map. For example, Fig. 1.11 shows with large red dots the first seven iterates for the initial conditions $(0.1, 0.1)$ with red lines and arrows indicating the temporal sequence. After only three or four iterations, the orbit is on the attractor to the resolution of the plot, and thereafter it dances around on the attractor forever, never revisiting a previous point. As a consequence, it is a good idea to discard the first few iterates (which constitute the *initial transient*)

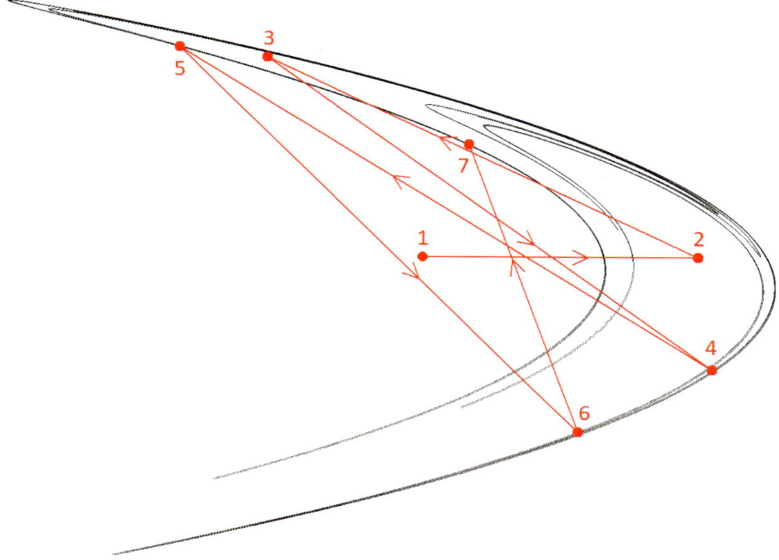

Fig. 1.11 The Hénon attractor showing the first seven iterates for an initial condition of $(0.1, 0.1)$ demonstrates how the orbit is rapidly drawn to the attractor.

when plotting an attractor since they are likely to lie off the attractor. If you are doing millions of iterations, it does not cost much to ignore the first hundred or first thousand iterates.

Figure 1.11 also gives you the chance to see the result of anti-aliasing since it is plotted with the same resolution as Fig. 1.6. Realize, that after a few million iterations, the orbit is visiting pixels that have already been visited, and so nothing new is revealed with additional iterations even though the values of the iterates never repeat exactly.

1.6 Chaos

A strange attractor is a type of fractal produced by a *chaotic* process in which the iterates are drawn to a region of negligible size, but they move around within that region, never repeating, in such a way that two nearby points separate at an exponential rate on average, eventually losing memory of where they have been. For that reason, it is impossible to predict where they will be at some time in the distant future without calculating all the intermediate values. Even that calculation is doomed to failure because of round-off errors arising from the fact that computers do their arithmetic with numbers having a finite number of digits (typically about 16). Chaotic systems are said to have *sensitive dependence on initial conditions*, commonly called the *butterfly effect*.

The finite precision of the numbers used by a digital computer means that the sequence of iterates will eventually repeat due to rounding, but you are not likely to ever see this occur in a two-dimensional map since the order of 10^{32} iterations are required to exhaust all the combinations of sixteen-digit numbers, and the square root of that number 10^{16} is an estimate of the average number of iterations until one of the previous iterates repeats. In higher dimensions, this *recurrence time* is longer yet.

To illustrate the chaos in the Hénon map, imagine starting with two nearby initial conditions for (x_0, y_0) of $(0.1, 0.1)$ and $(0.1001, 0.1)$ that differ by one part in a thousand. The next twenty-one x values and their absolute differences $|\delta x|$ are

0	0.100000000000000	0.100100000000000	$|\delta x| = 0.00100$
1	1.010000000000000	1.015971986000000	$|\delta x| = 0.00003$
2	−0.415158400000000	−0.415048706871498	$|\delta x| = 0.00011$
3	1.063500904073220	1.063619995094020	$|\delta x| = 0.00012$
4	−0.707995362150367	−0.708317103610763	$|\delta x| = 0.00032$
5	0.617289865264964	0.616687631553648	$|\delta x| = 0.00060$
6	0.254135902492518	0.255079780041023	$|\delta x| = 0.00094$
7	1.094767879869530	1.093914317606010	$|\delta x| = 0.00085$
8	−0.601682624363875	−0.598784013956475	$|\delta x| = 0.00290$
9	0.821599591314898	0.826215081764043	$|\delta x| = 0.00462$
10	−0.125541031137494	−0.135319110055053	$|\delta x| = 0.00978$
11	1.224415106695780	1.222228758364680	$|\delta x| = 0.00219$
12	−1.136531604248020	−1.131976125899670	$|\delta x| = 0.00456$
13	−0.441061190427662	−0.427249301940100	$|\delta x| = 0.01381$
14	0.386691555543542	0.404848409818428	$|\delta x| = 0.01816$
15	0.658339140091543	0.642362080512444	$|\delta x| = 0.01598$
16	0.509232873936001	0.543773863473142	$|\delta x| = 0.03454$
17	0.834456374171496	0.778742603718622	$|\delta x| = 0.05571$
18	0.177925445627184	0.314116099056900	$|\delta x| = 0.13619$
19	1.206016462369160	1.095486287954170	$|\delta x| = 0.11053$
20	−0.982888356819452	−0.585891460216787	$|\delta x| = 0.39700$
21	0.009307607951011	0.848069561969309	$|\delta x| = 0.83876$

and the y values follow along with a lag of one time step.

You will notice that at first the separation δx actually decreases. This is because the initial conditions are far from the attractor and become closer together as they are drawn to it. Furthermore, the initial perturbation was arbitrarily taken in the x direction, and that is not the direction in which the growth of their separation is greatest. You will also notice that δx does not increase monotonically, but that it sometimes decreases. However, a more precise calculation, averaging over many orbits on the attractor or following a single orbit for a very long time while continually readjusting the magnitude but not the direction of the perturbation to keep it small shows that the perturbation grows at an average exponential rate of 0.41922 for the Hénon map [Sprott (2003)]. This quantity is called the *Lyapunov exponent*, and it is a measure of how chaotic a system is, with negative values signifying that the orbit approaches a stable equilibrium (a fixed point) or cycles endlessly through a periodic sequence of values producing an attractor that consists of a small number of isolated points and thus is of little use for producing visually interesting fractal patterns.

A dynamical system has a *spectrum* of Lyapunov exponents as will be described in the next chapter, but the one of interest here is the largest, which determines whether the system is chaotic. The other Lyapunov exponents are useful for determining the dimension of the attractor as will be discussed later.

1.7   Basins of Attraction

It is not generally the case that every initial condition approaches an attractor. Rather each attractor is surrounded by a *basin of attraction*, which is something like the catch basin for a watershed. Initial conditions outside the basin of attraction typically approach infinity with repeated iterations, and we say such a system is *unbounded*, although it can also happen that there are multiple attractors, each with its own non-overlapping basin of attraction. Initial conditions on the basin boundary will cause the orbit to remain on the boundary forever, but even the slightest perturbation will make it approach the attractor or escape to infinity. The basin boundary is itself often a fractal, like the shoreline of a lake, and it is sometimes called a *repellor* since nearby points are pushed away from it, either toward the attractor or off to another attractor which may be far (even infinitely far) away.

The basin of attraction for the Hénon attractor is shown as cyan in Fig. 1.12. Its boundary is relatively simple and smooth, but it stretches

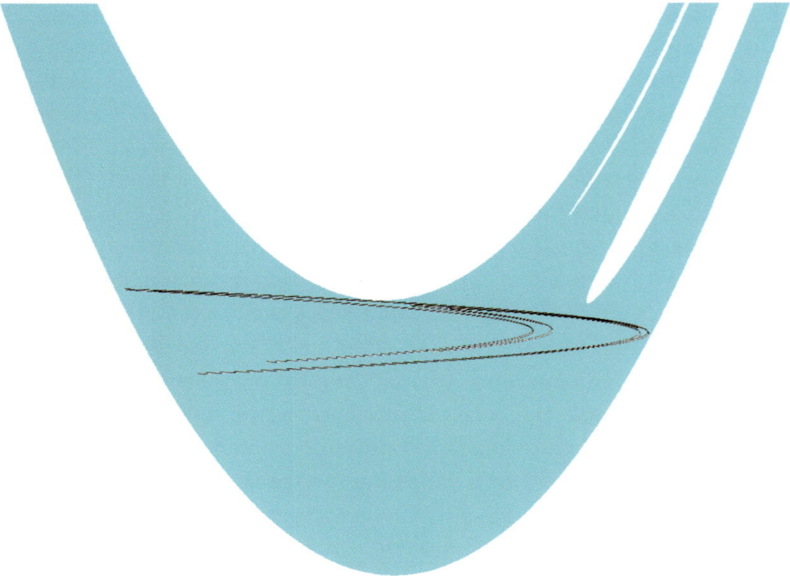

Fig. 1.12   The Hénon attractor (in black) is surrounded by a basin of attraction (in cyan) for which initial conditions go to the attractor, while those regions in white escape to infinity.

to infinity in some directions. The probability that a point a distance r from the Hénon attractor is in its basin decreases according to $P \approx 10.843/r^{1.855}$ in the limit of large r [Sprott and Xiong (2015)], and thus it is also a fractal but not a very elegant one. The attractor nearly touches its basin boundary as is common for chaotic systems, although there are *global attractors* for which the basin of attraction is the whole of space except for an infinite set of measure zero representing *unstable periodic points*. The white regions in Fig. 1.12 represent those initial conditions that are unbounded and approach infinity, and this region is sometimes called the *basin of infinity*.

1.8   A Third Dimension

Two-dimensional maps do not provide enough variety and are not sufficiently elegant for our purposes. Thus it is useful to consider a three-dimensional generalization given by

$$
\begin{aligned}
x_{n+1} &= f(x_n, y_n, z_n) \\
y_{n+1} &= g(x_n, y_n, z_n) \\
z_{n+1} &= h(x_n, y_n, z_n).
\end{aligned}
\tag{1.5}
$$

As an example, you can generalize the Hénon map by taking $f(x, y, z) = 1 - 1.4x^2 + 0.3z$, $g(x, y, z) = z$, and $h(x, y, z) = x$ giving

$$
\begin{aligned}
x_{n+1} &= 1 - 1.4x_n^2 + 0.3z_n \\
y_{n+1} &= z_n \\
z_{n+1} &= x_n,
\end{aligned}
\tag{1.6}
$$

which is just the standard Hénon map but with a transformation of variables (y is replaced by z) and an additional variable y that is the previous value of z or the second previous value of x. Since $f(x, y, z)$ and $h(x, y, z)$ do not depend on y, the dynamics is independent of y, and the map is still two-dimensional, but it provides a third variable to make a more interesting plot. You can recover the conventional Hénon map by plotting z on the vertical axis and x on the horizontal axis. You can also think of Eq. (1.6) as a *master-slave system* in which x and z comprise a master oscillator and y is a slave variable that obediently follows along without influencing the dynamics.

Now it is necessary to consider how to plot an object with a third variable on the computer screen or printed page, which are inherently two-dimensional. One way is to have the portion of the image that is closer to you appear in a darker shade of gray so that it occludes the portion that lies behind it. This can be done by first constructing a quantity z_p in the range 0 to 1 from $z_p = (z - z_{min})/(z_{max} - z_{min})$ and then assigning the following gray levels to the four pixels nearest to the point:

Pixel	Gray level
$(\lfloor x_p \rfloor, \lfloor y_p \rfloor)$	$255 - 255z_p \delta x_p \delta y_p$
$(\lfloor x_p \rfloor, \lfloor y_p \rfloor + 1)$	$255 - 255z_p \delta x_p (1 - \delta y_p)$
$(\lfloor x_p \rfloor + 1, \lfloor y_p \rfloor)$	$255 - 255z_p (1 - \delta x_p) \delta y_p$
$(\lfloor x_p \rfloor + 1, \lfloor y_p \rfloor + 1)$	$255 - 255z_p (1 - \delta x_p)(1 - \delta y_p).$

The factor of 255 can be altered to reduce the contrast and keep the most distant part of the image (where $z = z_{min}$ and $z_p = 0$) from fading into invisibility.

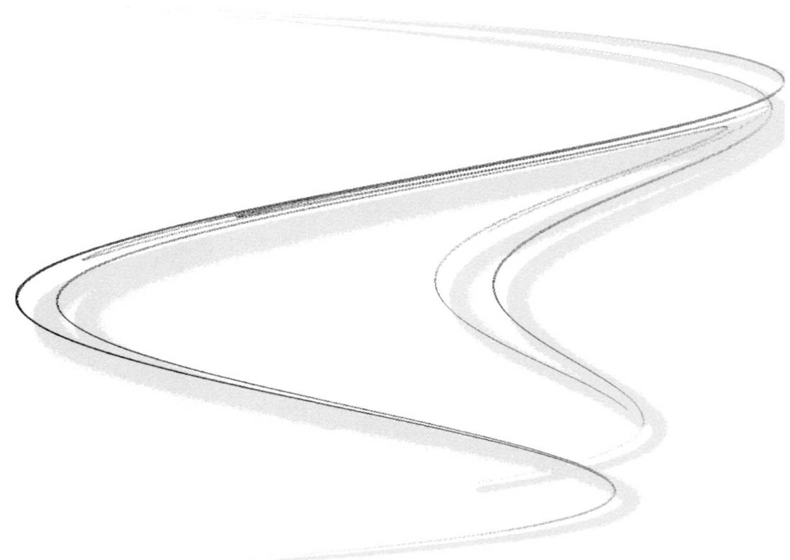

Fig. 1.13   The Hénon attractor embedded in a three-dimensional space from Eq. (1.6) with a shadow below and to the right of the attractor to enhance the illusion of a third dimension.

An additional embellishment that enhances the three-dimensional illusion is to add a subtle shadow below and to the right of each point in the image at a distance proportional to z_p as if the object were illuminated by a light over your left shoulder.

To simulate a diffuse light source, the shadow points are smeared out in a random Gaussian distribution G of mean 6.0 and variance 1.0 and plotted at a position (x_s, y_s) where $x_s = x_p + wz_p/5G$ and $y_s = y_p + hz_p/5G$ with a light gray level of 220. An easy way to generate a good approximation for G is to sum twelve uniform random deviates in the range of 0 to 1. Such uniform random numbers are provided by most computer languages, and more will be said about computer random number generators in Chapter 8. The resulting plot for the map in Eq. (1.6) is shown in Fig. 1.13.

Note that three types of dimension have been described. There is the dimension of the space in which the attractor is embedded, which is 3.0 in this example; there is the dimension of the system of equations (the number of variables that contribute to the dynamics), which is 2.0 for the Hénon map; and there is the dimension of the attractor, which is a fraction between

1.0 and 2.0 for the Hénon map. The embedding dimension and the system dimension are always integers, but the attractor dimension can take on any value up to the system dimension. An attractor can be embedded in a space of any dimension, and the attractor dimension will remain unchanged no matter how large the embedding dimension. You can even embed an attractor in a space smaller than the attractor dimension, in which case it is a projection onto a lower-dimensional space much like the shadow of a three-dimensional object on the two-dimensional ground.

There are other ways to display three-dimensional images. For example, the attractor can be plotted as an *anaglyph* in which two images are overlayed, one in red and the other in cyan, showing the attractor as seen separately by each eye, and then the image is viewed with special '3-D' colored glasses that allow each image to be viewed by the corresponding eye [Sprott (1992)].

1.9 Colorization

Figure 1.13 is still not what you would consider an 'elegant' fractal. At the very least, you need to add color. There are many ways to do this, but one straightforward method is to generalize Eq. (1.6) by adding a fourth variable u and then to map that new variable into some appropriately chosen color palette. The resulting map is given by

$$
\begin{aligned}
x_{n+1} &= 1 - 1.4x_n^2 + 0.3z_n \\
y_{n+1} &= z_n \\
z_{n+1} &= x_n \\
u_{n+1} &= y_n.
\end{aligned}
\tag{1.7}
$$

Suppose you want the color to change continuously from blue when $u = u_{min}$ to red when $u = u_{max}$. Define a quantity $u_p = (u - u_{min})/(u_{max} - u_{min})$ between 0 and 1, and then assign the red, green, and blue values according to $(r, g, b) = (255u_p, 0, 255 - 255u_p)$, respectively. Each component of the color is individually changed just as was done previously for the gray images where all three color components had the same value. The resulting color image is shown in Fig. 1.14.

But why stop there? Since there are three basic colors, you can extend the method to six dimensions with three of the variables (u, v, w) independently controlling the individual (r, g, b) values as described above. The

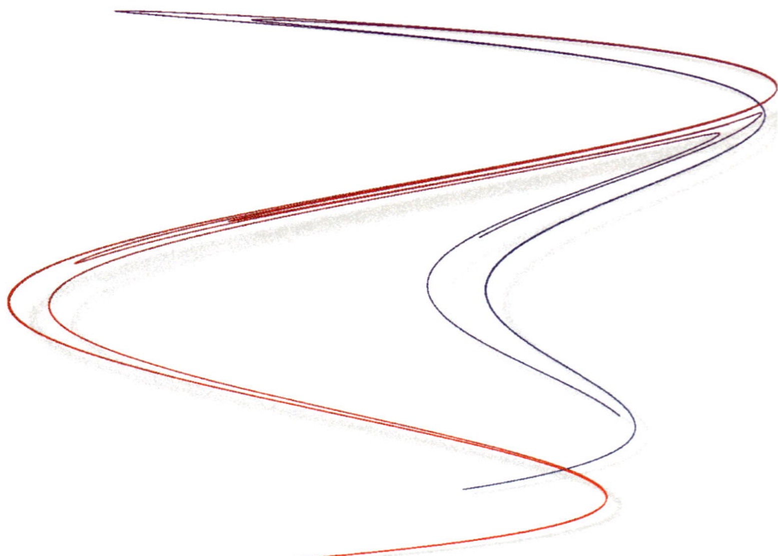

Fig. 1.14   Colorized Hénon attractor embedded in four dimensions from Eq. (1.7).

corresponding generalized Hénon map is

$$
\begin{aligned}
x_{n+1} &= 1 - 1.4x_n^2 + 0.3z_n \\
y_{n+1} &= z_n \\
z_{n+1} &= x_n \\
u_{n+1} &= y_n \\
v_{n+1} &= u_n \\
w_{n+1} &= v_n,
\end{aligned}
\tag{1.8}
$$

and an image produced in this way is in Fig. 1.15. This is the technique that will be used for most of the figures in this book, although we will consider some alternate coloring schemes in later chapters.

We are now in a position to consider more elegant examples.

1.10   Elegance

The Hénon map contains the function $f(x, z) = 1 - ax^2 + bz$, where a and b are parameters chosen somewhat arbitrarily by Hénon as $(1.4, 0.3)$ to give a chaotic solution, but they were not in any sense optimized, especially for the elegance of the resulting attractor. It turns out that the dimension of the Hénon attractor depends sensitively on the parameters and has a maximum

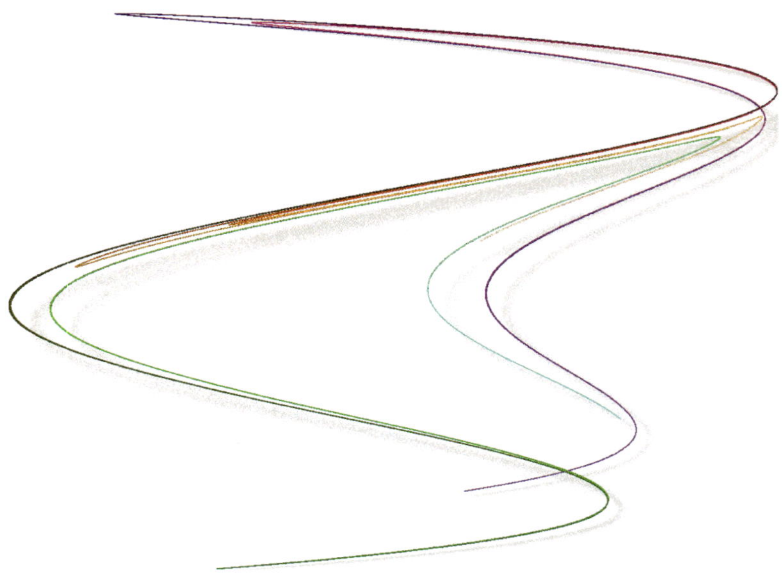

Fig. 1.15   Colorized Hénon attractor embedded in six dimensions from Eq. (1.8).

value of about 1.3075 when $a = 1$ and $b = 0.54272$ [Sprott (2007)] and a corresponding attractor as shown in Fig. 1.16. While this dimension is only about 4% larger than the case in Fig. 1.15, the resulting attractor is arguably more elegant, suggesting that the fractal dimension might be a sensitive factor in determining the elegance of an image, and this will turn out to be the case.

While elegance is purely subjective, most values of the parameters do not give chaotic solutions and their corresponding strange attractors, and thus they are usually rather inelegant. In fact, you only need to change the parameters in Fig. 1.16 to $(1, 0.543)$ (a change of 0.05%) to get unbounded solutions and no attractor at all. On the other hand the parameters $(1, 0.5)$ (a change of 8%) give an attractor that consists of only eight dots, a so-called *period-8 cycle* with dimension zero. Thus it is reasonable to infer that the search for strange attractors, especially elegant ones, is a difficult task, and every elegant solution will likely have inelegant solutions nearby in the space of parameters.

If you concede that Fig. 1.16 is more elegant than Fig. 1.15, perhaps there is a way to make it even more elegant. One possibility is to plot an iterate of x greater than the second on the vertical axis. For example,

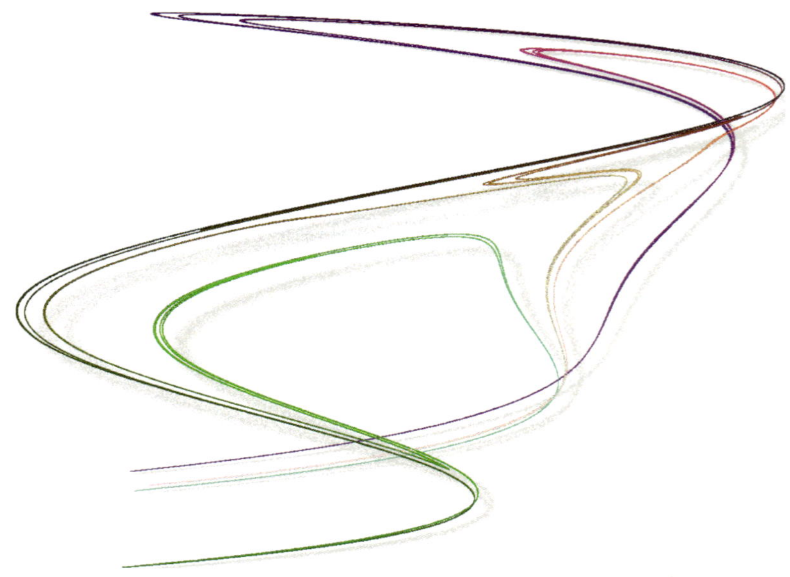

Fig. 1.16   Colorized Hénon attractor from Eq. (1.8) but with parameters $a = 1$ and $b = 0.54272$ where the attractor has its largest fractal dimension.

consider the mapping

$$\begin{aligned}
x_{n+1} &= 1 - x_n^2 + 0.54272 z_n \\
y_{n+1} &= w_n \\
z_{n+1} &= x_n \\
u_{n+1} &= y_n \\
v_{n+1} &= z_n \\
w_{n+1} &= v_n,
\end{aligned} \qquad (1.9)$$

where now y is the fourth previous value of x rather than the second as in Eq. (1.8). The resulting image in Fig. 1.17 is at least more complex if not more elegant than Fig. 1.16. This is despite the fact that both cases have identical fractal dimensions (1.3075). The extra complexity comes from repeatedly stretching and folding the attractor back onto itself as is typical for a chaotic process.

The mapping of the six variables into position and colors amounts to 90-degree rotations of the attractor in the six-dimensional embedding space. There are in fact $6! = 720$ ways to do this, each giving a different image, but with many looking rather similar. (The quantity $6!$ is called *six factorial* and is given by $6! = 6 \times 5 \times 4 \times 3 \times 2 = 720$.) For example, exchanging

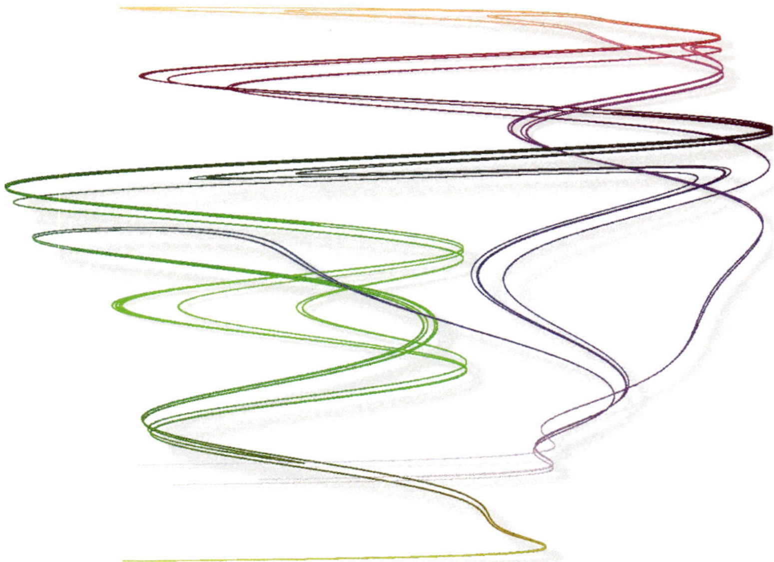

Fig. 1.17   Colorized Hénon attractor from Eq. (1.9) in which the fourth rather than the second previous iterates is plotted.

x and y only amounts to a 90-degree rotation of the image in the viewing plane, and any two choices for the horizontal and vertical coordinates that differ by the same time delay will look the same except for the depth and colors. Of course you can embed the attractor in spaces higher than six, giving even more possibilities, although it is ultimately necessary to project it onto two dimensions to make the plot.

For example, what would happen if you were to plot the twentieth previous iterate on the vertical axis? This is easily done by storing the values of x in a circular buffer of size 20 and using the oldest such value as y before overwriting it with the current value of x. The resulting image as shown in Fig. 1.18 is perhaps elegant in some minimalist sense, but you would probably agree that it is less elegant than the previous cases, and so there is a point of diminishing returns when increasing the time-delay embedding dimension. Continuing the process to the hundredth iterate or beyond fills in the entire rectangle with what looks like a uniform gray color as would occur if the iterates were generated from random numbers.

What has happened is that the sensitive dependence on initial conditions that characterizes all chaotic processes has destroyed the correlation

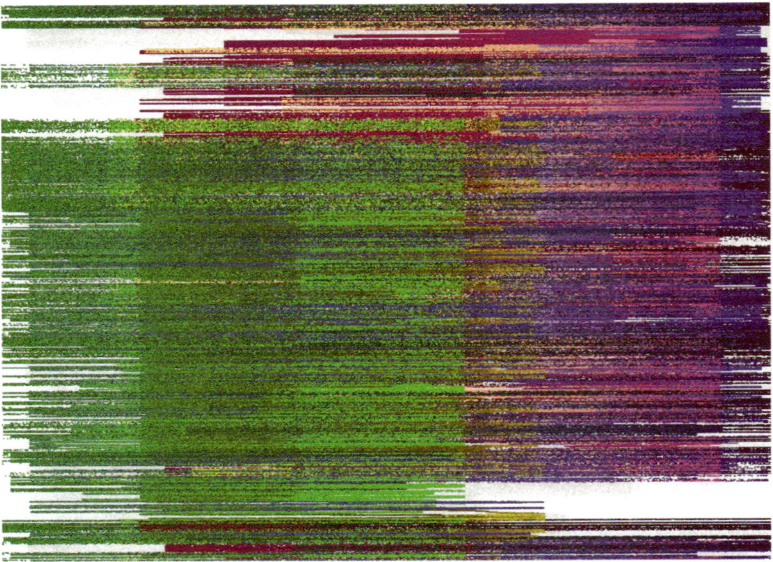

Fig. 1.18   Colorized Hénon attractor from Eq. (1.9) but with the twentieth previous iterate plotted on the vertical axis degrades the correlation between x and y due to sensitive dependence on initial conditions.

between x and y, and thus their values are effectively random. This is despite the fact that the fractal dimension is still only 1.3075. A physical analogy is taking a thin sheet of paper whose dimension is 2.0 and wadding it into a ball so that it becomes effectively space-filling and looks three-dimensional on the large scale even though it is still essentially a two-dimensional object.

In any event, it should be clear that the fractal dimension is not a sufficient measure of elegance since very different images can have identical dimensions. Rather we need some quantity that is a measure of how visually space-filling the image is. A simple way to construct such a measure is to count the number of unique pixels visited by an orbit on the attractor after a fixed large number of iterations, remembering that a single point will be spread over four pixels because of the anti-aliasing.

To make the quantity independent of the resolution of the image, it is better to express it as a fraction of the total number of pixels available to the image, ignoring the shadow. The calculated values of the pixel fraction are $PF = 0.087$ for Fig. 1.17 and $PF = 0.997$ for Fig. 1.18. Presumably the most elegant images will tend to have intermediate values of PF. That is to say, they will have some white space, but not too much. We will return to estimate an optimal value for PF in the next chapter.

1.11 The Lozi Map

To show another example, consider the Lozi map [Lozi (1978)] with parameters chosen to give the largest fractal dimension [Sprott (2007)] and mapped into position and color according to

$$
\begin{aligned}
x_{n+1} &= 1 - 1.7052|x_n| + 0.5896z_n \\
y_{n+1} &= w_n \\
z_{n+1} &= x_n \\
u_{n+1} &= y_n \\
v_{n+1} &= z_n \\
w_{n+1} &= v_n.
\end{aligned}
\tag{1.10}
$$

This mapping can be considered as a *piecewise-linear* variant of the Hénon map, making it amenable to analytic calculations. The resulting image in Fig. 1.19 has a fractal dimension of 1.4042 and a pixel fraction of $PF = 0.200$. Strange attractors from piecewise-linear maps tend to be less elegant than those from quadratic maps because of their straight lines and angular corners.

1.12 Three-dimensional Maps

All the previous examples involved two-dimensional maps but embedded in some higher-dimensional space. As a result, their fractal dimension is relatively low, and their appearance is somewhat line-like, although the line can be folded back onto itself many times giving the impression of a two-dimensional object. More interesting fractals require maps of higher dimension. For example, a three-dimensional map with a strange attractor is given by

$$
\begin{aligned}
x_{n+1} &= 1 - 0.3x_n^2 - 0.5y_n + 0.7z_n \\
y_{n+1} &= x_n \\
z_{n+1} &= y_n \\
u_{n+1} &= z_n \\
v_{n+1} &= u_n \\
w_{n+1} &= v_n.
\end{aligned}
\tag{1.11}
$$

The resulting image in Fig. 1.20 has a fractal dimension of 1.8907 and a pixel fraction of $PF = 0.349$. Since the dimension is less than 2.0, the

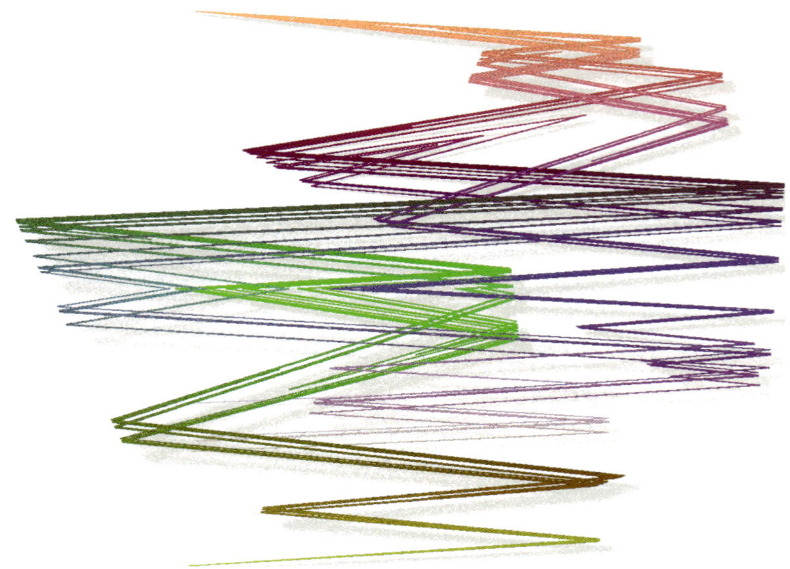

Fig. 1.19   Colorized Lozi attractor from Eq. (1.10).

object is still mostly holes. but it begins to look more like a surface than a tangled line, and we finally have an example of a fractal that you might unequivocally call 'elegant.' You can extend the method to any known chaotic mapping or to one that you might discover using the search methods in the next chapter.

1.13   Resolution and Image Size

Most of the images in this book are printed in a size of 4 inches (1200 pixels) wide by 3 inches (900 pixels) high so that two images will fit comfortably on a page with captions and at a resolution of 300 dots per inch to speed the calculation and limit the size of the resulting graphics files. However, the program is written to facilitate changing the resolution and image size. To give an example of the improvement that is possible, Fig. 1.21 shows the same image as Fig. 1.20 but elongated to 6.5 inches vertically and printed at 600 dots per inch (2400 pixels wide by 3900 pixels high), which requires 8.67 times as many iterations to produce the equivalent image. This image required about 15 minutes to compute on a typical desktop computer versus about two minutes for Fig. 1.20, and it does show some additional detail

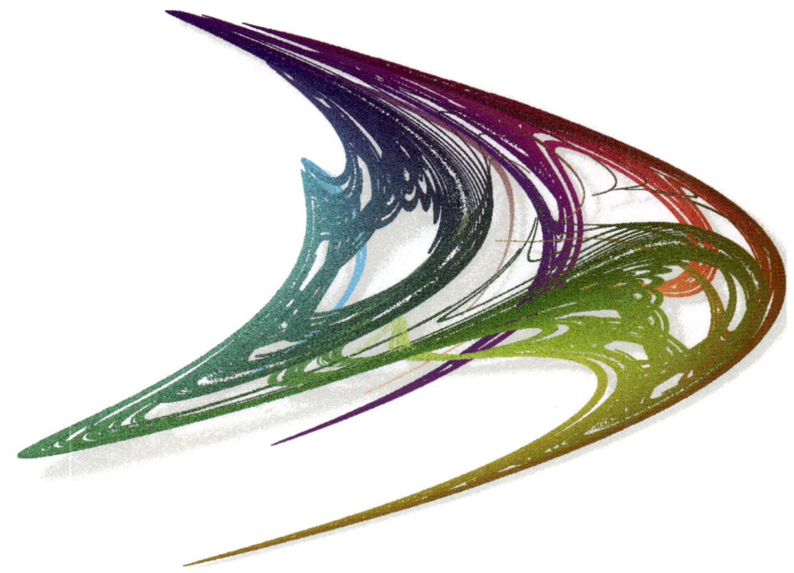

Fig. 1.20   Colorized three-dimensional attractor from Eq. (1.11).

that is most evident only with magnification. The compressed **png** graphics file is also about eight time larger (6 MB). It will be left for you to judge whether the increased elegance justifies the required additional computation and storage.

1.14   A Comment on Elegance

Since elegance is inherently subjective, you might want to decide whether you consider Fig. 1.17 or 1.19 to be the more elegant of the two and why. More generally, consider the following set of ten familiar symbols:

$$1\ 2\ 3\ 4\ 5\ 6\ 7\ 8\ 9\ 0$$

Ignoring the fact that they have mathematical meaning and perhaps even emotional content, decide which of the above symbols is the most visually elegant and which is the least elegant, or, better yet, rank order them according to their elegance. There is no 'correct' answer to this exercise, but it will cause you to contemplate your understanding of what it means

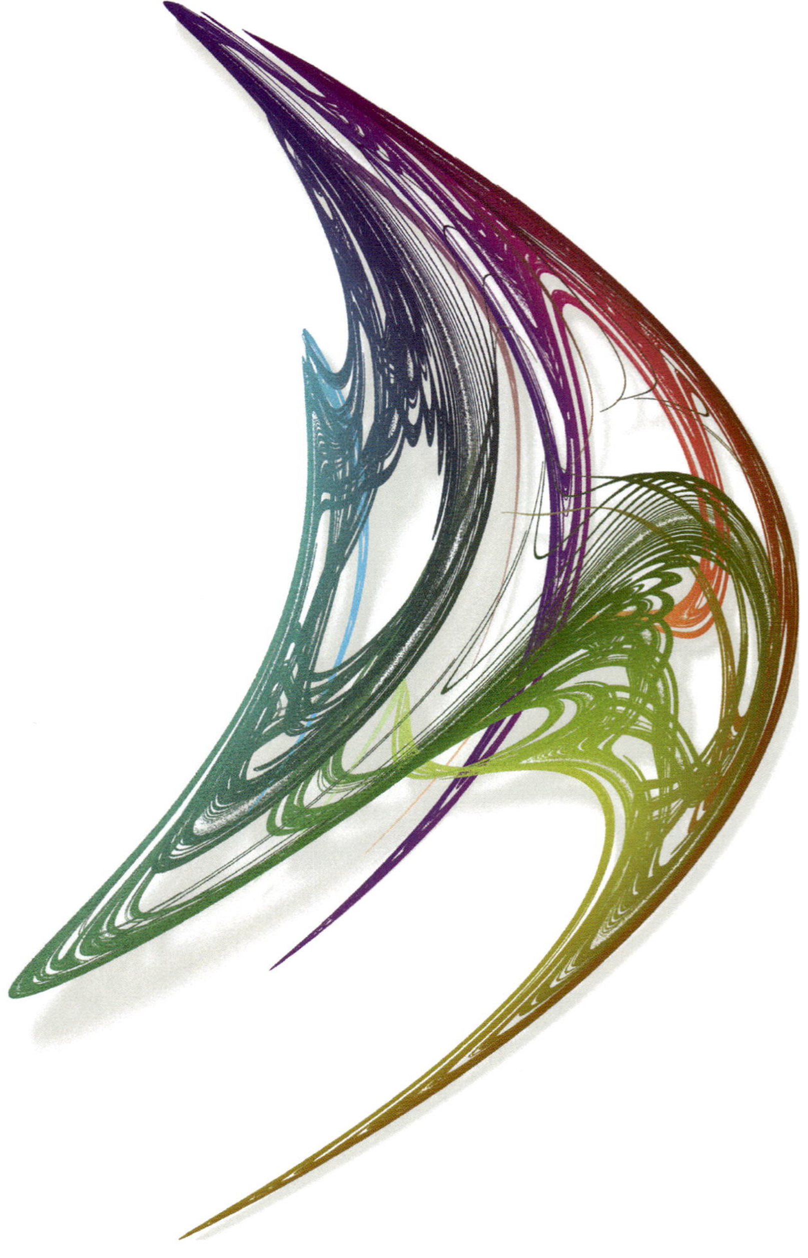

Fig. 1.21   The same image as in Fig. 1.20 but elongated and at higher resolution.

to be elegant and how one might automate the selection of elegant fractals from among the vast majority that are unremarkable.

There are at least three qualities that distinguish these symbols: symmetry, smoothness, and simplicity. Perhaps you can decide whether they factor into your evaluation, and if so how, and whether there are other qualities that you would add to the list.

If you found this simple exercise interesting, try the same for the following set of twenty-six equally familiar symbols:

$$A\ B\ C\ D\ E\ F\ G\ H\ I\ J\ K\ L\ M$$
$$N\ O\ P\ Q\ R\ S\ T\ U\ V\ W\ X\ Y\ Z$$

Do other fonts lead to different conclusions? What do you consider to be the most elegant font? Here is a sans-serif (Gothic) font:

$$A\ B\ C\ D\ E\ F\ G\ H\ I\ J\ K\ L\ M$$
$$N\ O\ P\ Q\ R\ S\ T\ U\ V\ W\ X\ Y\ Z$$

Do you consider it more or less elegant than the previous Times Roman font? Keep your conclusions in mind as you explore and examine the fractal images in the following chapters.

Chapter 2

Search Techniques

This chapter will describe efficient methods for finding new examples of elegant attractors among the much more numerous unbounded and inelegant solutions that occur. It will introduce some additional nomenclature and quantitative measures used throughout the book and will show examples of attractors found using those methods for a particular six-dimensional quadratic map. Finally, it will suggest ways to automate the selection of those attractors most likely to appeal to a human.

2.1 Six-dimensional Quadratic Maps

The two- and three-dimensional maps with two or three parameters used as examples in the previous chapter are not sufficiently general to permit a wide variety of elegant attractors. Thus we generalize Eq. (1.11) to six dimensions with all possible linear and squared terms:

Case **A:**

$$
\begin{aligned}
x_{n+1} &= a_1 x + a_2 x^2 + a_3 y + a_4 y^2 + a_5 z + a_6 z^2 \\
&\quad + a_7 u + a_8 u^2 + a_9 v + a_{10} v^2 + a_{11} w + a_{12} w^2 \\
y_{n+1} &= x \\
z_{n+1} &= y \\
u_{n+1} &= z \\
v_{n+1} &= u \\
w_{n+1} &= v.
\end{aligned}
\tag{2.1}
$$

In this and the equations to follow, the subscript n has been omitted from the variables on the right-hand side of the equations to simplify the notation and improve readability. Equation (2.1) is still a highly restricted example of a six-dimensional map since all the nonlinearities are in the x equation

31

and only quadratic nonlinearities are allowed, but with twelve adjustable parameters $a_1, a_2, \ldots, a_{12}$, it permits an enormous variety of solutions. The parameters are like numbers on a combination lock, which open the door to an elegant fractal image when chosen properly. Most combinations produce images that are completely inelegant, containing just a few dots or orbits that quickly leave the region of interest.

You can, of course, imagine maps with many more than twelve parameters that would provide an even wider range of possible images [Sprott (1993b)], but it turns out that more is not necessarily better. Keeping the equations relatively simple adds a kind of mathematical elegance to the visual elegance of the images, and it speeds the search, simplifies the notation, and reduces the amount of information required to reproduce the images. As you will soon see, there is no lack of variety, and it will be left for you to explore the many possible more complicated maps.

You may wonder why the constant term 1.0 that was present in all the maps in the previous chapter has been omitted in Eq. (2.1). The reason is that it can be removed without loss of generality through a linear transformation of the variables. For example, the Hénon map in Eq. (1.2) can be written as

$$
\begin{aligned}
u_{n+1} &= -1.4u^2 - 1.767792536u + 0.3v \\
v_{n+1} &= u,
\end{aligned}
\tag{2.2}
$$

where $u = x - 0.631354477$ and $v = y - 0.631354477$. Omitting the constant keeps the orbit closer to the origin and ensures that the origin is a fixed point, albeit usually unstable (a *repellor*) in the sense that nearby initial conditions are repelled from it rather than attracted to it. You might want to check the algebra or plot the attractor for Eq. (2.2) and convince yourself that it is a displaced but otherwise identical version of Fig. 1.6.

2.2 Parameter Range

In principle, the twelve parameters can take on any values between $-\infty$ and $+\infty$. However, values too close to zero lead mostly to solutions that settle to a small number of isolated points, and ones that are too large usually lead to unbounded solutions, neither of which will produce an elegant attractor. Furthermore, two systems whose parameters differ only slightly will tend to look similar, but certainly not always.

Therefore, it is better to choose the parameters within some intermediate range between $-a_{max}$ and $+a_{max}$ and to discretize the values within that range. To choose an appropriate value for a_{max}, consider the following table in which the percentages of the different types of solutions are shown for various values of a_{max} with parameters chosen uniformly and randomly over the corresponding range:

a_{max}	Fixed Point	Periodic	Chaotic	Unbounded
0.125	49.20%	50.80%	0.00%	0.00%
0.250	35.10%	64.37%	0.20%	0.33%
0.500	18.35%	44.57%	10.58%	26.50%
1.000	1.49%	4.02%	3.78%	90.71%
2.000	0.09%	0.07%	0.09%	99.75%
4.000	0.02%	0.00%	0.00%	99.98%
8.000	0.00%	0.00%	0.00%	100.00%

For the purposes of this table, 'fixed point' means the attractor is a single point (also called a *period-1 orbit*), 'periodic' means the period of the orbit is between two and six, and 'chaotic' includes all cases that do not fall into one of the other three categories and thus is a candidate for an elegant attractor, although not necessarily strictly chaotic since it could also be a periodic attractor with a period greater than six or a so-called *chaotic transient*.

From the above table, it appears that the optimum value for a_{max} is approximately 0.5 in the sense that it gives the greatest chance that an arbitrarily chosen system will have an elegant attractor (approximately 10%). However, limiting the range to that small a value risks missing interesting cases, and there is little penalty for choosing a slightly larger value of a_{max} since the ubiquitous unbounded solutions can be quickly identified and eliminated. Thus a value of 1.0 is a better choice, while a value of 2.0 or higher unnecessarily degrades the efficiency of the search since nearly all the solutions are unbounded.

While there is a continuum of values for each parameter in the range $-a_{max}$ to $+a_{max}$, it is better to use discrete values, say in increments of 0.1, to avoid two cases being overly similar and so that the resulting equations are more elegant and easily written. For $a_{max} = 1.0$, this gives 21 different values for each parameter $(-1.0, -0.9, \ldots, 0.9, 1.0)$. It is convenient to expand the range slightly and use the 26 letters of the alphabet to represent values in the range -1.2 to 1.3 according to the following table:

A	−1.2		N	0.1
B	−1.1		O	0.2
C	−1.0		P	0.3
D	−0.9		Q	0.4
E	−0.8		R	0.5
F	−0.7		S	0.6
G	−0.6		T	0.7
H	−0.5		U	0.8
I	−0.4		V	0.9
J	−0.3		W	1.0
K	−0.2		X	1.1
L	−0.1		Y	1.2
M	0.0		Z	1.3

This scheme has the virtue that each system in the form of Eq. (2.1) can be described by a single twelve-character string, which allows the cases to be stored efficiently and reproduced by typing a simple twelve-character code. Since there will be other cases later, this string will be prefaced by the letter A to denote that it is a system of the type given by Eq. (2.1). You can view this string as the 'name' of the attractor, but do not try to pronounce it!

For values chosen in this range, approximately 97% of the solutions are unbounded, 0.5% are fixed points, 1.5% are periodic with a period less than seven, and the remaining 1% have potentially elegant attractors. It is simple to program the computer to recognize and discard the 99% that are unbounded or periodic with a small period.

You might object that twenty-six values of each parameter is an insufficient number to identify all the elegant solutions. However, since there are twelve parameters, there are $26^{12} \approx 10^{17}$ different combinations. Even if only 1% of those are strange attractors, there are still about 10^{15} different cases. To put this in perspective, if you viewed them at a rate of one per second, it would take you about thirty million years to see them all. Thus the number of cases is effectively infinite, and like snowflakes, no two that you are likely to produce in a random search are exactly alike, and every one you produce is unlikely ever to have been seen before. Thus you are truly in uncharted territory when using this procedure, and, like a biologist exploring the wild, you have a chance to stumble across something new and important.

You might also wonder if the spacing of 0.1 for the parameter values is reasonable. In fact, changing any one parameter by an amount 0.1 more often than not gives a qualitatively different solution, often an unbounded one. A reason is that strange attractors are usually located close to their basin boundary as the Hénon map example in Fig. 1.12 shows. Sometimes such a small change gives an attractor that resembles the original one, but only rarely are the two cases indistinguishable. If you find an example that you especially like, you might try changing each of the parameters slightly to see if you can produce one that is even better. In fact, try it for the images in this book, since that was not done when they were selected.

2.3 Sample Images

To test the proposed method, several million instances of Eq. (2.1) were calculated with randomly chosen parameters, producing several thousand images, constrained only by requiring that the pixel fraction (PF) covered by the attractor be greater than 0.03 to eliminate cases that are too sparse. Figures 2.1 to 2.4 are four images chosen to show a range of attractor dimensions. For these and all the other similar images in this book, the number of iterations has been taken as $10wh = 10.8$ million for $w = 1200$ and $h = 900$, so that each pixel is visited ten times on average, but with most never visited.

Figure 2.1 shows a case that is relatively sparse with a pixel fraction of $PF = 0.146$. This case is actually periodic with a period of 31 512, but the points appear as short line segments as a result of the very slow approach to the attractor and the fact that only the first 10^5 iterates were discarded before plotting. You can demonstrate this by discarding more initial iterates, which causes the line segments to collapse to 31 512 individual dots. Consequently, this case is not chaotic, nor is it a fractal, but rather it is a periodic cycle with an attractor whose dimension is $DF = 0$.

Figure 2.2 shows another relatively sparse case with an even smaller pixel fraction of $PF = 0.051$ but that consists of a continuous curve with many wiggles. This is an example of *quasiperiodicity*, which means that the period is an irrational number so that each iteration visits a point on the curve at a different phase of the cycle. The iterates do not move sequentially along the line, but they appear to visit random spots until the line eventually fills in. The resulting plot is called a *drift ring*. It is an attractor with a dimension of $DF = 1.0$, and thus it is not a fractal. Zooming in on the line would not reveal any additional small-scale structure.

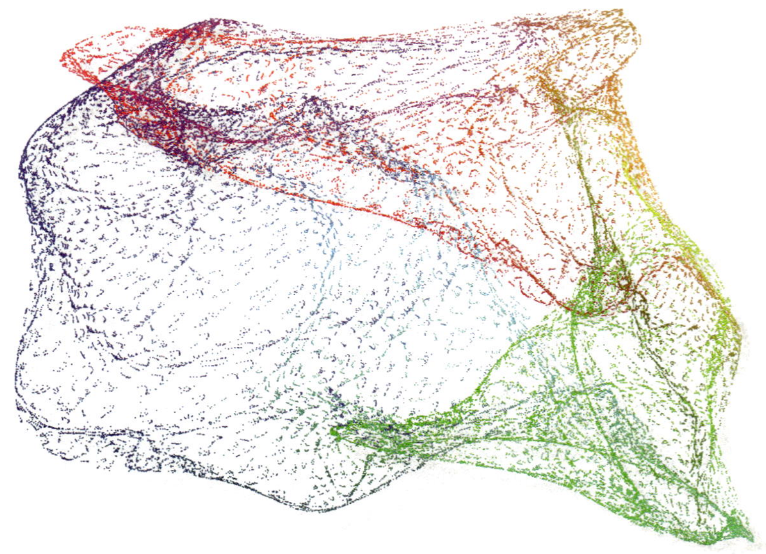

Fig. 2.1   `AJOOVNJGIHBQQ` $(PF = 0.146, CP = 0.023, LE = -0.0001, DF = 0)$.

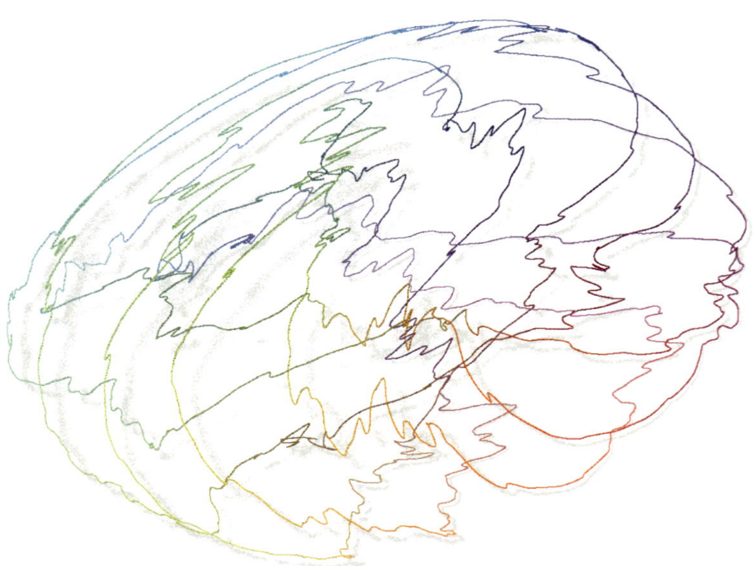

Fig. 2.2   `AXYAEWJIYGOOX` $(PF = 0.051, CP = 0.015, LE = 0, DF = 1)$.

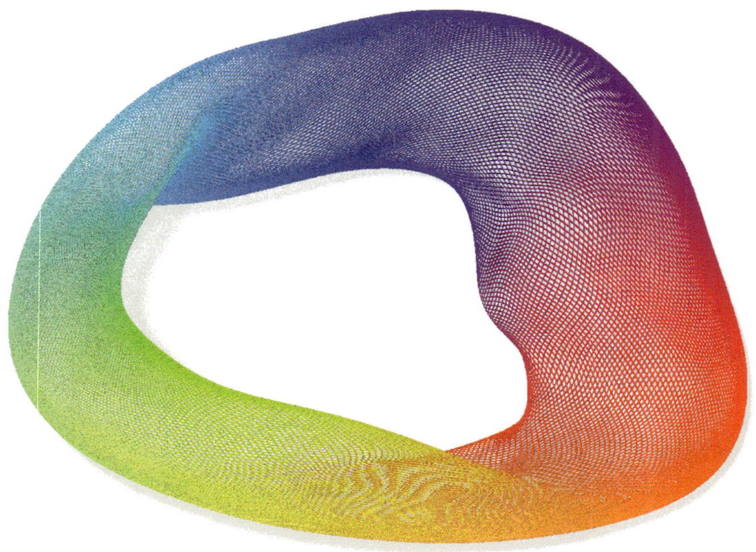

Fig. 2.3   `AQNDWSPNHJUKM` ($PF = 0.451, CP = 0.299, LE = 0, DF = 2$).

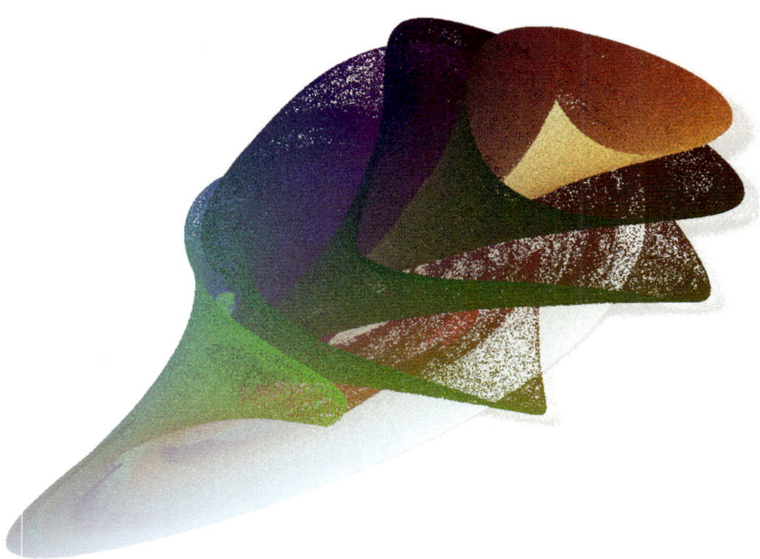

Fig. 2.4   `AUCPMIXORIJKR` ($PF = 0.471, CP = 0.906, LE = 0.0207, DF = 2.2323$).

Figure 2.3 shows a case where the drift ring wraps around on the surface of a *torus* (a doughnut) never quite closing on itself but eventually filling the entire two-dimensional surface. It consists of two periods that are not rational ratios of one another (they are said to be *incommensurate*), nor are they a rational fraction of the iteration step size. This object is not a fractal either, but it is an attractor with a dimension of $DF = 2.0$ (a surface). About 2×10^9 iterations are required for the surface of the torus to completely fill in at the resolution of the plot.

Attractors with integer dimensions greater than 2.0 are possible, but more commonly, strange attractors occur, such as the example in Fig. 2.4. This case is a fractal with a dimension of $DF = 2.2323$. If you were to examine it closely, you would see that the curled up leaves actually consist of additional closely packed leaves like a head of lettuce. The resulting image is a fractal and surely is at least somewhat elegant.

2.4   Cluster Probability

The pixel fraction (PF) and fractal dimension (DF) are two quantities that characterize the previous images, and it is evident that elegant attractors occur for a wide range of their values. Thus it is useful to have additional quantities that can be easily calculated and that relate to the appearance of the attractor and thus to its elegance.

One such quantity is the *cluster probability* (CP), which has been used to specify the connectedness of forest landscapes [Sprott *et al.* (2002)]. It is defined as the probability that an arbitrary pixel on the attractor has all four of it nearest neighbor pixels (north, south, east, and west) also on the attractor. The cluster probability would be zero if the attractor consisted of isolated points or thin non-overlapping lines, and would be equal to 1.0 if it fills the entire surface densely. The values of CP are given along with PF in the figure captions.

The cluster probability and pixel fractal are not completely independent. If the pixel fraction is small, the cluster probability will necessarily also be small, and if the pixel fraction is large, the cluster probability will also tend to be large. However, the two quantities are not proportional. In particular, Figs. 2.5 and 2.6 show two cases with similar pixel fractions $(PF \approx 0.22)$ but with very different cluster probabilities, and Figs. 2.7 and 2.8 show two cases with similar cluster probabilities $(CP \approx 0.087)$ but with very different pixel fractions. Apparently neither quantity alone is sufficient to characterize an image.

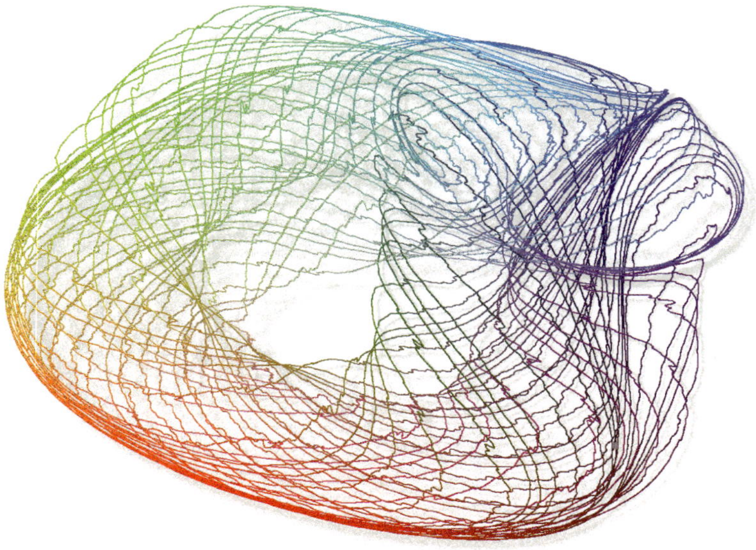

Fig. 2.5 `AONMMHQTMOCKV` ($PF = 0.214, CP = 0.068, LE = 0, DF = 1$).

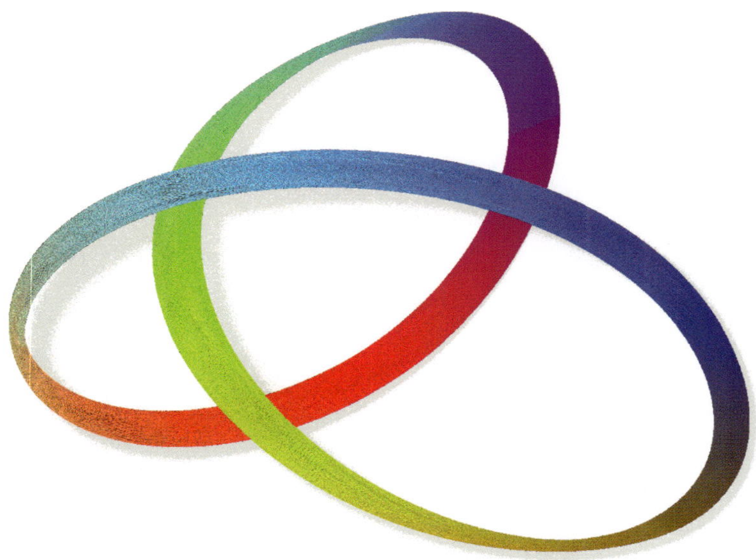

Fig. 2.6 `AJRIMPTPNVJKO` ($PF = 0.220, CP = 0.968, LE = 0, DF = 2$).

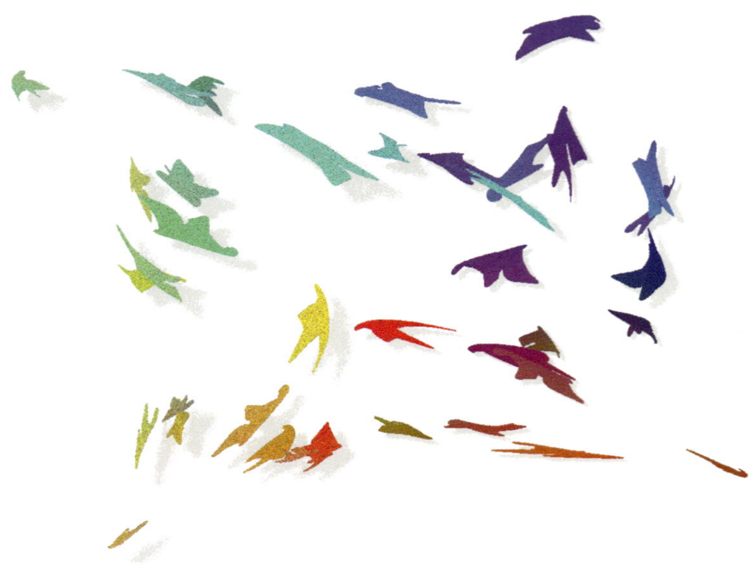

Fig. 2.7   **AHDDSCWKELVIW** ($PF = 0.080, CP = 0.871, LE = 0.0041, DF = 2.1626$).

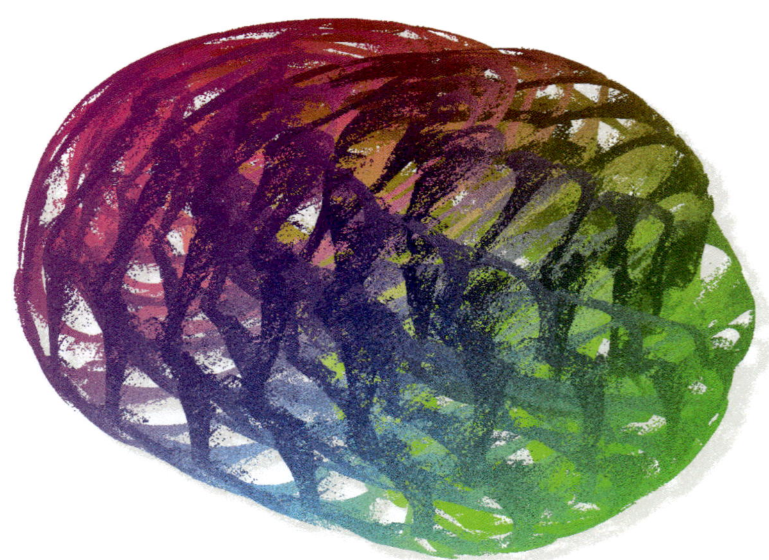

Fig. 2.8   **AWFPEULKFNVKF** ($PF = 0.666, CP = 0.879, LE = 0.0005, DF = 2.1585$).

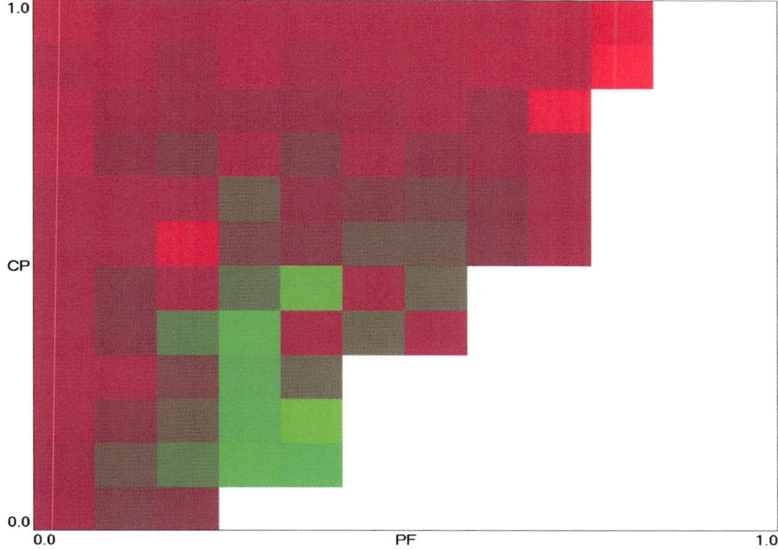

Fig. 2.9   Regions of elegance are shown in green and inelegance in red as a function of the pixel fraction (PF) and cluster probability (CP) shows a 'sweet spot' in the vicinity of $PF = CP = 1/3$.

2.5   Aesthetic Evaluation

The combination of pixel fractal and cluster probability provides a useful indication of how likely an attractor is to be elegant. For a collection of 12 462 images produced using randomly chosen coefficients, 175 were selected as being especially elegant, and another 144 were selected as being especially inelegant (acknowledging that such judgements are highly subjective, but these are extreme cases for which you would likely concur). All the other cases were unremarkable and might be judged differently by different people. The cases were grouped into a 12×12 array of boxes in the space of (PF, CP), and the net elegance $(good - bad)/total$ was plotted in Fig. 2.9 as a color for those boxes that had at least four cases, with green indicating most elegant and red indicating least elegant. The figure clearly shows a 'sweet spot' in the vicinity of $PF = CP = 1/3$. Thus the most elegant fractals have about two thirds white space and are neither too solid nor too broken into small pieces.

This result means that you can refine the search for elegant fractals by giving preference to the relatively few that lie in the vicinity of the optimum values of PF and CP. For example, you could eliminate any cases for which

$$(CP - 1/3)^2 + (PF - 1/3)^2 > (1 + 8r)/9 \qquad (2.3)$$

where r is a uniform random number in the range of 0 to 1. This does not significantly speed the search for elegant fractals since the inelegant cases must be calculated to determine their values of PF and CP, but it greatly reduces the number of images that you must examine to find the gems among the junk.

For example, using Eq. (2.3) in combination with checking for periodicity and unbounded solutions eliminates 99.9% of the cases but still produces about 15 candidates per hour on a 2018-era desktop personal computer, most of which are at least somewhat elegant. Thus you can run the program (available at `http://sprott.physics.wisc.edu/fractals/elegant/elegfrac.exe`) overnight and have upwards of a hundred cases to examine and admire the next morning.

However, if you restrict the cases to a range of PF and CP that is too narrow, that will limit the variety of cases produced, and you will miss some unusual ones, which is why they are retained but with a low probability. Figures 2.10 and 2.11 show two cases in the vicinity of $PF = CP = 1/3$. Figure 2.11 is a long-duration transient to a periodic cycle with a period of 211 253.

2.6 Machine Learning

While it is possible to use the above procedure to optimize the search for elegant fractals, it is laborious since you must first determine values of PF, CP, DF and perhaps other quantities that correlate with the elegance of the image. You would probably prefer to automate the process and bypass making a plot such as Fig. 2.9. One way to do that is to train a *feed-forward artificial neural network* on the data that produced that figure and use the output of the network to predict the probability that a particular combination of PF and CP will be elegant. Such a network could be very simple with two inputs, one for PF and the other for CP, a hidden layer of neurons, and a single output that is a measure of the likely relative elegance of the image. You can then set a threshold on that quantity or save cases randomly in proportion to its value.

There is a certain mystique about artificial neural networks because they are crudely modeled on the brain. However, in this case, the method just amounts to choosing a general mathematical model with many parameters that are adjusted optimally to fit the training data. Conventionally,

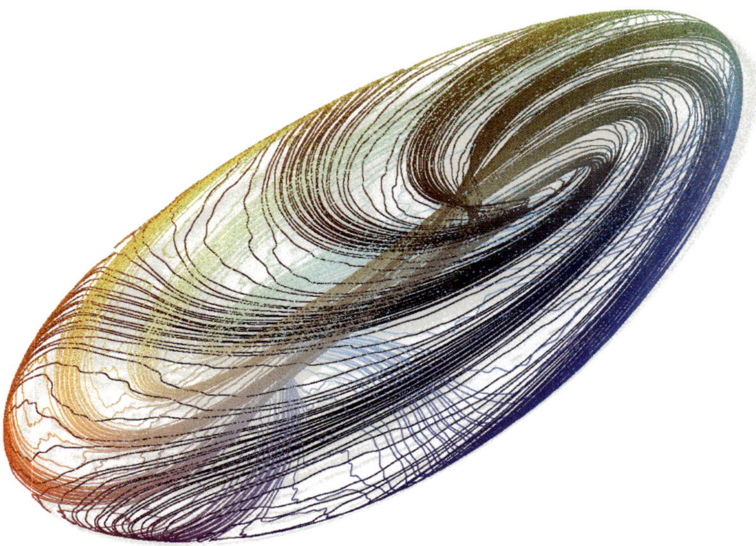

Fig. 2.10   AYRCHRFRHKNRG $(PF = 0.349, CP = 0.318, LE = 0, DF = 1)$.

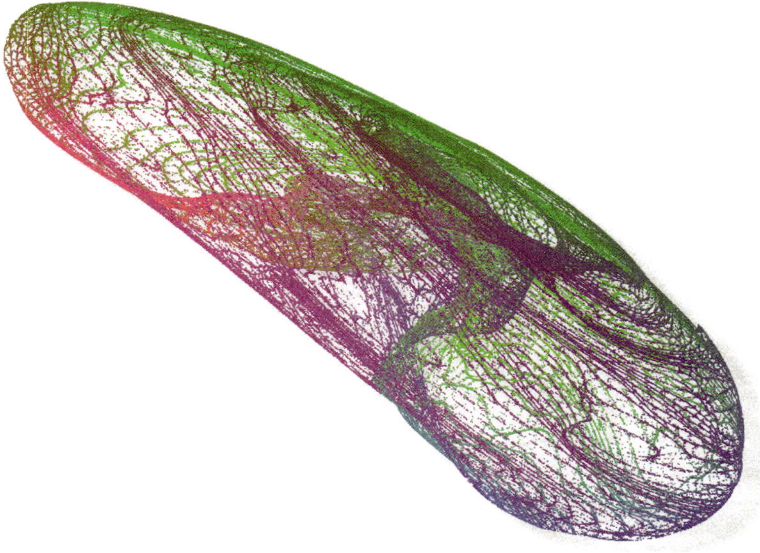

Fig. 2.11   AKIQMREJPPRYA $(PF = 0.299, CP = 0.346, LE = -0.000002, DF = 0)$.

the neurons are taken to be sigmoidal functions such as the hyperbolic tangent. Such a simple network with sufficiently many neurons is a *universal approximator* in the sense that it can represent any continuous function to arbitrary precision [Hornik *et al.* (1989)]. This feature is less impressive than it sounds because polynomials are also universal approximators, but a sigmoid is a better basis function for modeling real-world data because the output is bounded so that it behaves better when extrapolated beyond the range where data exist, whereas polynomials often given unbounded solutions.

To illustrate the method, the 67 data points that were used to color the squares in Fig. 2.9 were taken as input to the function

$$P = \sum_{i=1}^{N} a_i \tanh(b_i + c_i PF + d_i CP) \tag{2.4}$$

where N is the number of neurons and (a_i, b_i, c_i, d_i) is a $4 \times N$ matrix of parameters that are adjusted to give a least squares fit to the data. There are sophisticated ways to optimize the parameters such as *backpropagation* [Schmidhuber (2015)], but this network is sufficiently simple that a trial-and-error approach suffices. Choose the parameters in Eq. (2.4) randomly from a Gaussian distribution centered on the best case found so far, and gradually shrink the width of the Gaussian as better cases are found. The solution converges slowly, but the method is easily implemented and helps avoid getting trapped in a local minimum. This method is somewhat akin to *simulated annealing* [Press *et al.* (2007)].

The choice of how many neurons to use is a tradeoff of speed and accuracy. However, be careful to make the number of parameters $(4N)$ smaller than the number of data points (67) to avoid over-fitting. You want a smooth approximation to the data that does not attempt to fit the statistical noise. Using $N = 8$ gives the parameters

i	a_i	b_i	c_i	d_i
1	7.0873	-1.8008	-3.5305	8.3015
2	6.6309	1.6300	0.8703	-6.6892
3	4.3592	-1.3197	1.2124	3.8015
4	3.1819	2.0481	5.6590	-10.2534
5	2.1642	-1.2390	-1.5118	-3.8365
6	1.8316	3.0205	-7.7117	5.2168
7	0.8786	6.5257	0.0695	-14.0852
8	0.4256	4.0525	-0.8425	-5.6293

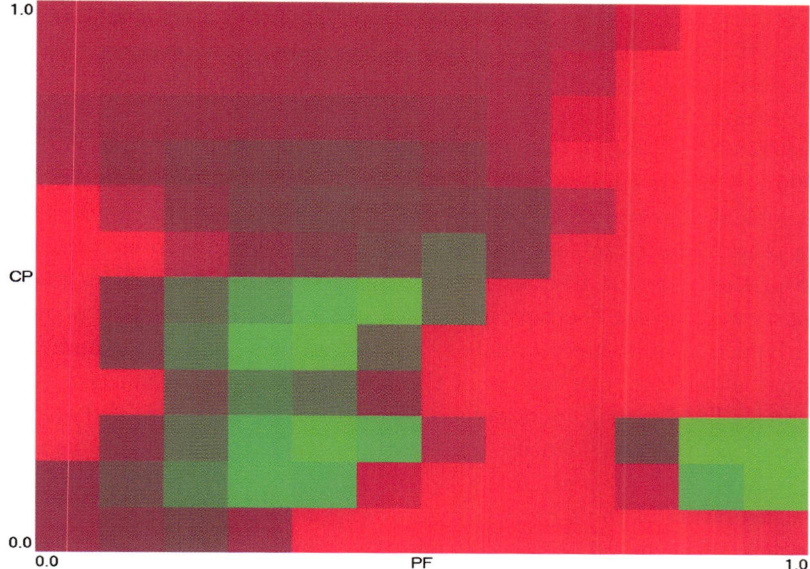

Fig. 2.12 Prediction based on a neural network approximation to the data in Fig. 2.9 reproduces the 'sweet spot' near $P = CP = 1/3$ but also shows an island of elegance at large PF and small CP where no data exist.

and the resulting prediction from Eq. (2.4) is shown in Fig. 2.12 with the same coloring as Fig. 2.9. The neural network gives a plausible prediction outside the range where data occur, including an island of elegance at large PF and small CP where examples are absent in the data and unlikely to exist.

For this example, the advantage of training a neural network is not very evident, but in a case with more than two quantities describing the image, it gives you a general method to automate the process of selecting ones that you are likely to find appealing.

2.7 Largest Lyapunov Exponent

An example of an additional quantity that might correlate with the elegance of an attractor is the *largest Lyapunov exponent* (LE), which is a measure of how chaotic a system is [Sprott (2003)]. Dynamical systems have a spectrum of Lyapunov exponents, but we consider here only the largest (most positive) of them. Fixed points and periodic cycles give negative values of LE, drift rings and tori give zero values, and strange attractors give positive values. Only those cases with $LE > 0$ are fractals.

Previous studies have shown that large values of LE are judged as less aesthetically appealing than small values [Sprott (1993a,b); Aks and Sprott (1996)], which is curious since the LE is a dynamic quantity rather than a topological one, and thus one would not expect its effect to be discernible in a static image. In fact, many of the cases in this chapter have $LE = 0$, and they are not fractals despite their elegance. It may be that the Lyapunov exponent is acting as a surrogate for some other hidden variable with which it is highly correlated. If your interest is in chaos and fractal strange attractors, calculation of the LE is probably the quickest way to eliminate fixed points, periodic cycles, drift rings, and tori, and that was the method used to identify the equations with chaotic solutions in the companion book *Elegant Chaos* [Sprott (2010)].

Calculation of the largest Lyapunov exponent is relatively simple. You calculate two orbits separated by some tiny amount. One orbit, the unperturbed one, is the one that produced the previous figures. The perturbed orbit (sometimes called a 'satellite orbit') has an initial condition displaced from the unperturbed one by a small distance δr_0 in an arbitrary direction. The choice of δr_0 is not critical, but a good value is the order of the square root of the precision of the numbers. For example, the IEEE 754 standard for double precision floating point numbers [Goldberg (1991)] specifies that they store 52 bits for the mantissa of the number and 11 bits for the exponent. Thus a good choice would be $\delta r_0 = 2^{-26} \approx 1.5 \times 10^{-8}$. This value allows the separation to shrink or expand by at least eight orders of magnitude before causing an overflow or underflow.

After each iteration, the new separation of the orbits δr is calculated. The *local* largest LE is given by $\log(r/r_0)$. Usually base-e logarithms are used, but sometimes you will see LEs quoted as 'bits per iteration,' which indicates that the logarithm is base-2. Then the satellite orbit is readjusted so that its separation is given by δr_0 but without altering the direction of the perturbation, and the process is repeated to get a succession of local LEs.

The *global* largest LE is the average along the orbit of the local LEs after infinitely many iterations. Since it is impractical to iterate infinitely many times, we instead iterate several million times and assume the *finite time Lyapunov exponent* is an adequate approximation to the global LE. You may want to discard the first few dozen or more iterates to give the perturbation a chance to rotate into the direction of the maximum growth (or minimum decay).

The local LE will usually vary enormously along the orbit, and so the average will converge slowly, and it will always be difficult to distinguish a small positive LE from one that is zero. Consequently, it is wise to display a running average so that you can judge when the value has converged. As a rule of thumb, if you want an LE that is accurate to d digits, you will need the order of 10^{2d} iterations, or a hundred million iterations for four-digit precision. The value of the calculated LE is included in the figure captions.

Applying the calculation to the cases that produced Fig. 2.9 gives an average LE of 0.0035 ± 0.0075 for those cases that were deemed elegant and 0.0221 ± 0.0211 for the ones that were deemed inelegant. Even the elegant cases with $LE > 0.0001$ have an average value of 0.0071 ± 0.0094, which is considerably smaller than the inelegant cases. Thus it is reasonable to conclude that weakly chaotic attractors are more elegant than strongly chaotic ones, perhaps reflecting the fact that some chaos in nature adds a degree of interest and unpredictability, while too much is perceived as random and totally disordered.

The local LE provides an alternate method for coloring an attractor. For example, points where it is positive could be colored in red, and points where it is negative could be colored in blue. Alternately, you could choose a continuum of colors from pure blue to pure red as the value of the local LE increases. No such examples are shown here, but they exist in the chaos literature [Hoover *et al.* (2015)].

For the images shown in this book, the Lyapunov exponent was not used as a selection criterion because its calculation would slow the computation by a factor of two, whereas PF and CP can be quickly calculated after the image is produced and provide much the same information. However, the LE values have been calculated to high accuracy (four significant digits) in a lengthy computation for those cases that were selected by other methods.

2.8 Spectrum of Lyapunov Exponents

The preceding discussion pertains to the *largest* Lyapunov exponent, and justifiably so since that is the quantity that determines whether a system is chaotic and how chaotic it is. However, for a system with six dimensions such as Eq. (2.1), there are actually six Lyapunov exponents, which are conventionally ordered from the largest (or least negative) to the smallest (or most negative). This *spectrum of Lyapunov exponents* is useful for distinguishing between drift rings and tori and for determining the dimension of the attractor.

Several methods are available for calculating the *LE* spectrum [Geist *et al.* (1990)], but they are all somewhat involved since they usually require repeated reorthonormalization of the expansion and contraction vectors for an ellipsoid of initial conditions. The method used here is patterned after Wolf *et al.* (1985), whose very readable paper actually includes two methods, one for when the equations are known and another for when all you have is a time series of data. The first method is the one employed here with the modification that the *Jacobian matrix* (the matrix of partial derivatives) for the map is determined numerically rather than analytically. Calculation of a reliable Lyapunov exponent spectrum from a time series is a difficult proposition that is not recommended unless the equations that produced the data are unknown and thus no other method is available.

Applying the algorithm to the figures in this chapter gives the following results for the six Lyapunov exponents for each figure:

Fig. 2.1	−0.0001	−0.0003	−0.0140	−0.0324	−0.3566	−0.8775
Fig. 2.2	0.0000	−0.0032	−0.0141	−0.0419	−0.3769	−0.5306
Fig. 2.3	0.0000	0.0000	−0.0830	−0.3158	−0.4657	−0.7450
Fig. 2.4	0.0207	0.0065	−0.1169	−0.1598	−0.3752	−1.0410
Fig. 2.5	0.0000	−0.0001	−0.0072	−0.1350	−0.4807	−0.6613
Fig. 2.6	0.0000	0.0000	−0.0012	−0.0371	−0.0638	−1.8242
Fig. 2.7	0.0041	−0.0001	−0.0244	−0.0783	−0.1943	−0.4622
Fig. 2.8	0.0005	−0.0000	−0.0031	−0.1802	−0.2365	−0.2417
Fig. 2.10	0.0000	−0.0001	−0.0559	−0.4232	−0.4258	−0.4747
Fig. 2.11	−0.0000	−0.0000	−0.0090	−0.1246	−0.1551	−1.0636

The largest Lyapunov exponent for each case agrees with the value calculated using the simpler method in the previous section and shown in the figure captions. Only three of the cases are chaotic as indicated by their positive largest *LE*. Figure 2.4 is an example of *hyperchaos* because it has two positive Lyapunov exponents, which means that there are two orthogonal directions in the six-dimensional space for which nearby orbits diverge exponentially. Hyperchaotic attractors do not look much different from ordinary chaotic ones (those with a single positive Lyapunov exponent).

Five of the cases have a largest *LE* of zero, and three of those have two zero Lyapunov exponents. Those with one zero exponent are the drift rings, and those with two are the tori. Note also that some of the Lyapunov exponents are very small and negative (-0.0000 and -0.0001) and can be easily mistaken as zero if you truncate the calculation too soon.

The sum of the positive Lyapunov exponents is one measure of the *entropy* of the system [Pesin (1977)], and it might offer a possible criteria for elegance. The *sum* of *all* the Lyapunov exponents is a measure of the exponential rate at which a six-dimensional hypervolume of initial conditions contracts. All the cases here have a negative value of this sum, ensuring that they are all attractors with an attractor dimension less than 6.0. Such systems are said to be *dissipative*. If the sum of the *LE*s were zero, the system is said to be *conservative* in the sense that the hypervolume occupied by a cluster of initial conditions is conserved in time, although it can stretch in some directions and simultaneously contract in other directions like a turbulent incompressible fluid, leading to chaotic solutions but not an attractor. If the sum of the *LE*s were positive, the hypervolume is forever expanding, which can only happen if the orbits are unbounded, and thus such cases are automatically excluded in the search process.

You should note that if the search had been limited to chaotic cases (those whose largest *LE* is positive), many elegant images would have been excluded (those for which the attractor is not strange and hence not fractal). For some purposes, that would be desirable, but the non-chaotic cases are shown by way of comparison. Often it is difficult to distinguish a strange attractor from a complicated torus or a transiently chaotic orbit based solely on the appearance of the orbit, and the value of the *LE* is the only definitive indicator.

2.9 Kaplan–Yorke Dimension

One of the main uses for the spectrum of Lyapunov exponents is to provide an estimate of the dimension of the attractor. If the largest *LE* is zero, then the attractor has an integer dimension with a value equal to the number of zero Lyapunov exponents. Thus one zero *LE* means the attractor is a one-dimensional drift ring, and two zero *LE*s means the attractor is a two-dimensional torus. It is also possible to have higher-dimensional tori, but no such examples have been shown because they are not common and their attractor when projected onto a two-dimensional surface tends not to be very elegant since it is too filled-in ($PF \to 1.0$).

The more interesting case is where the largest *LE* is positive and the orbit is chaotic, giving rise to a strange attractor with a fractional dimension.

There are many ways to estimate the dimension of a strange attractor, and they all give somewhat different values. However, Kaplan and Yorke (1979) proposed a simple and easily reproducible method based on the spectrum of Lyapunov exponents.

The idea is to begin summing the LEs, denoted by λ_i starting with the most positive one λ_1, but stopping before the sum becomes negative. The number d of LEs in the sum is the integer part of the fractal dimension. Then the fractional part is determined by interpolation of the zero crossing when the next (negative) LE is added to the sum,

$$DF = d + \frac{1}{|\lambda_{d+1}|} \sum_{i=1}^{d} \lambda_i. \tag{2.5}$$

The result DF is the *Kaplan–Yorke dimension*, and its value is shown in the figure captions.

You will note that all the previous examples of strange attractors in this chapter have a fractal dimension greater than 2.0. For iterated maps, the fractal dimension can take on any value less than the dimension of the system (6 in this case). In fact, it has been previously shown [Sprott (1993a,b); Aks and Sprott (1996)] that some of the most elegant examples of strange attractors have a dimension in the range of $1 < DF < 2$. Thus we provide five such examples in Figs. 2.13 to 2.17 with the following Lyapunov exponents:

Fig. 2.13	0.0018	−0.0051	−0.0415	−0.0644	−0.1848	−1.3179
Fig. 2.14	0.0005	−0.0009	−0.0026	−0.1570	−0.2089	−0.2908
Fig. 2.15	0.0077	−0.0104	−0.0207	−0.0209	−0.0281	−0.4846
Fig. 2.16	0.0093	−0.0096	−0.0394	−0.0998	−0.2404	−0.9376
Fig. 2.17	0.0001	−0.0004	−0.0224	−0.0360	−0.0623	−0.2357

The image in Fig. 2.17 has been enlarged and plotted at twice the resolution of the other cases.

Fractals usually (but not always) have a fractional dimension, and thus many of the previous examples are not fractals (strange attractors) despite their elegance. For periodic cycles, drift rings, and tori, zooming into the image will not reveal additional small-scale structure, while zooming into a strange attractor will show structure on all scales, but you will usually just find closely spaced lines that are made up of even more closely spaced lines rather than a miniature version of the entire strange attractor.

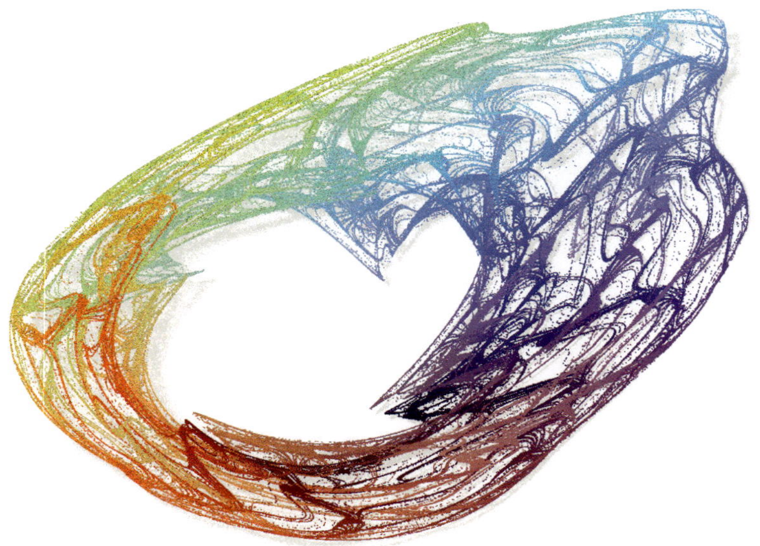

Fig. 2.13   **ATWBOIKTRCONQ** $(PF = 0.316, CP = 0.388, LE = 0.0018, DF = 1.3602)$.

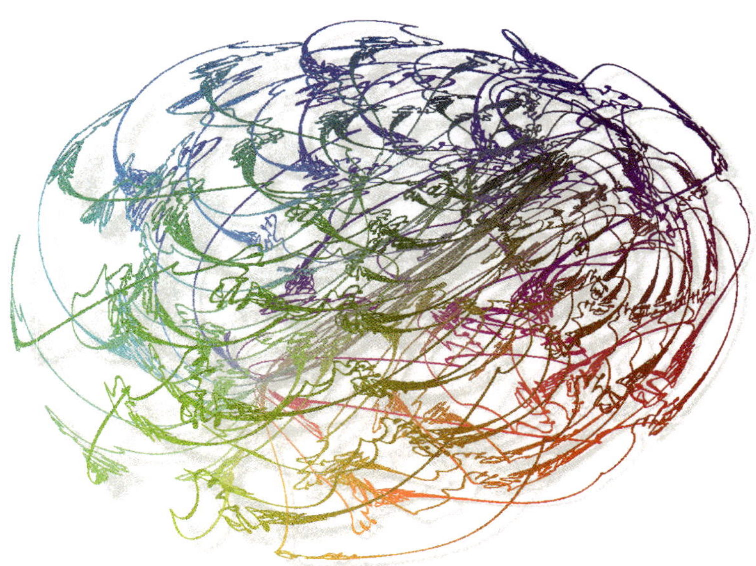

Fig. 2.14   **ARNJIZJISRHKF** $(PF = 0.256, CP = 0.431, LE = 0.005, DF = 1.5422)$.

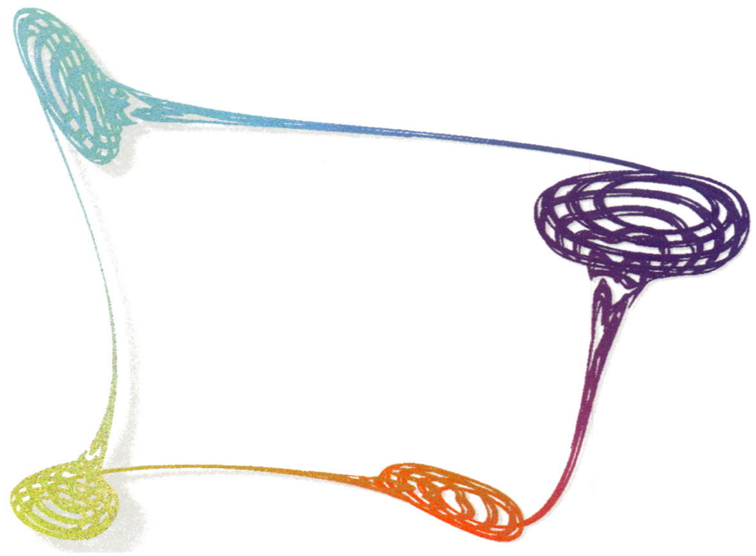

Fig. 2.15   **AHLAUCHDRIMDU** ($PF = 0.102, CP = 0.726, LE = 0.0077, DF = 1.7357$).

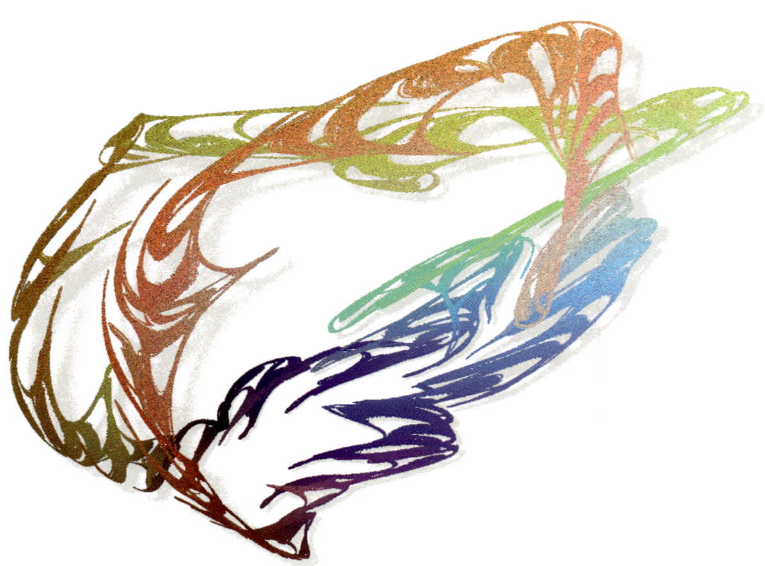

Fig. 2.16   **AIXITBXEYGQLG** ($PF = 0.240, CP = 0.757, LE = 0.0093, DF = 1.9700$).

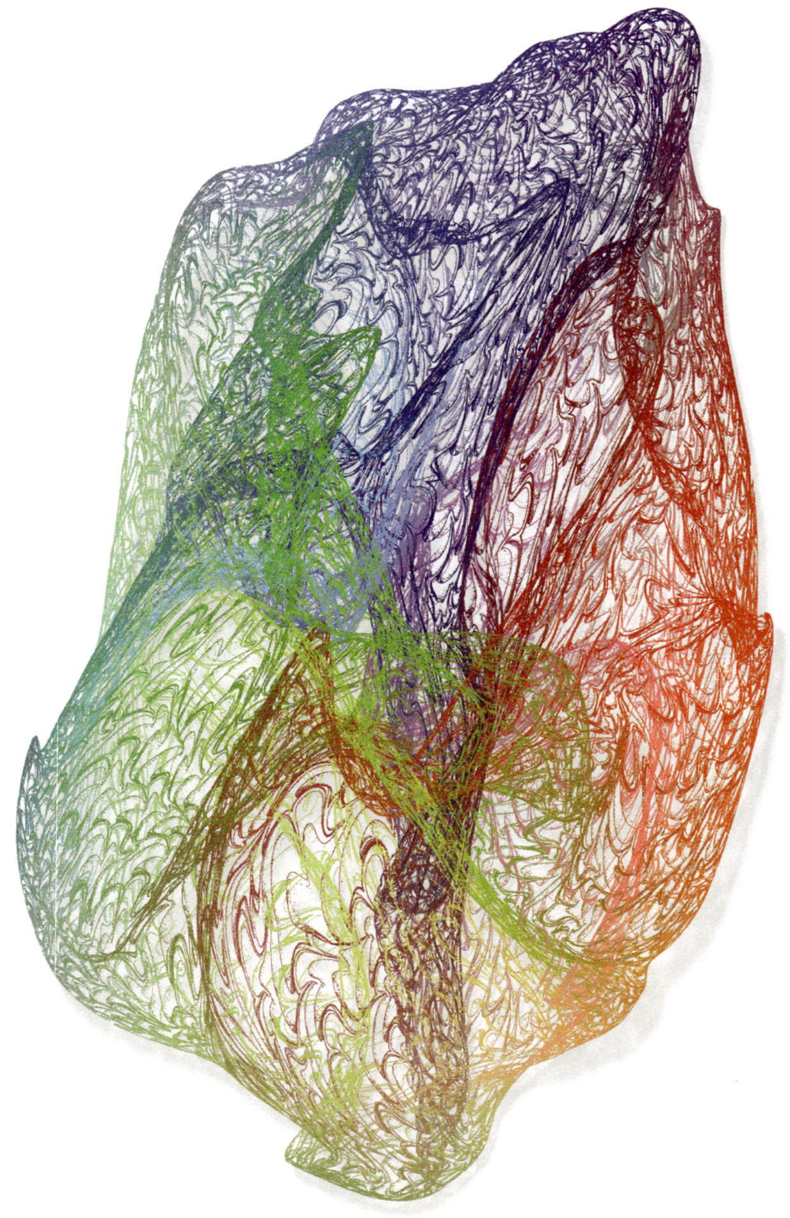

Fig. 2.17   AMSHOQYLKJXFM ($PF = 0.395, CP = 0.425, LE = 0.0001, DF = 1.3310$).

     Elegant Fractals

2.10   Correlation Dimension

There are other simple ways to estimate the dimension of an attractor, especially where you only have the data that produced the image rather than the equations from which they came. One of the most common is the *correlation dimension* [Grassberger and Procaccia (1983)]. The idea is as follows. Choose two arbitrary (but different) points on the attractor, and calculate the Euclidean distance between them (the square root of the sum of the squares of their separation along each dimension of the space). Do this for many random pairs of points, and calculate the probability $P(\epsilon)$ that two arbitrary points are separated by a distance less than ϵ. From a plot of $\log P(\epsilon)$ versus $\log \epsilon$, there will typically be a range of ϵ at small values of ϵ for which the curve is a straight line whose slope is the correlation dimension,

$$D2 = \lim_{\epsilon \to 0} \frac{d \log P(\epsilon)}{d \log \epsilon}. \tag{2.6}$$

It is impossible to take the limit all the way to $\epsilon = 0$ because you will run out of data pairs when ϵ becomes too small, and the curve will have large statistical fluctuations. Thus there is a 'scaling region' that must be chosen appropriately, usually by visual inspection of the curve, although there are ways to automate the process [Sprott and Rowlands (2001)].

The correlation dimension is denoted by $D2$ to distinguish it from the Kaplan–Yorke dimension DF and because it is one member of a whole spectrum of *generalized dimensions* [Rényi (1970)]. The '2' comes from the fact that it involves comparing pairs of points. Other common dimensions are the *capacity dimension* $D0$ and the *information dimension* $D1$ [Sprott (2003)]. Most strange attractors are actually *multifractals* [Harte (2001)], which means that the various dimensions are different with $D0 \geq D1 \geq D2 \geq D3 \geq \ldots$, and usually none of them will agree precisely with DF whose value is often closest to $D1$.

The correlation dimension has the virtue that it is the easiest of the generalized dimensions to calculate and it gives double weight to those regions of the attractor that are visited most often by the orbit so that it is arguably a more meaningful measure than $D0$ which weights all regions of the attractor equally. It is also most useful when all you have is a time series of data points along the orbit such as when the data come from the observation of some dynamical process in nature as well as for fractals that are not produced by the iteration of a dynamical system.

Like the Lyapunov exponent, the correlation dimension can be calculated while the orbit is being computed. Suppose you maintain an array of

10^4 past data points along the orbit. Each time a new point is generated, its separation from a random point in the array is calculated, and with a probability of 10^{-4} that point is replaced by the most recent point. Thus each iteration of the map generates one pair of points, and the separations of all such pairs are used to increment bins having the corresponding value of the integer part of $\log_2 \epsilon$, from which $P(\epsilon)$ can be estimated. Then you can calculate $D2$ from Eq. (2.6) using the slope of the curve over the range where the number of pairs with separation less than ϵ is between say 10^3 and 10^6. The value of 10^3 ensures reasonable statistics, and 10^6 assumes that meaningful values begin to appear only after a million iterations with a well-defined slope over three orders of magnitude of $P(\epsilon)$.

Unfortunately, the correlation dimension converges slowly compared with the Kaplan–Yorke dimension, and the two are only approximately equal as the following table with about 3×10^{10} iterations for each case shows for the previous examples in this chapter:

Figure	DF	$D2$
Fig. 2.1	0	0.155
Fig. 2.2	1	1.014
Fig. 2.3	2	1.976
Fig. 2.4	2.2323	1.987
Fig. 2.5	1	1.018
Fig. 2.6	2	2.016
Fig. 2.7	2.1626	1.911
Fig. 2.8	2.1585	1.869
Fig. 2.10	1	1.070
Fig. 2.11	0	0.231
Fig. 2.13	1.3602	1.196
Fig. 2.14	1.5422	1.368
Fig. 2.15	1.7357	1.521
Fig. 2.16	1.9700	1.580
Fig. 2.17	1.3310	1.171

Note that correlation dimension $D2$ tends to be less than the Kaplan–Yorke dimension DF as expected for a multifractal since $DF \approx D1 \geq D2$. The correlation dimension has not been used as part of the selection criterion for the images in this book, although you could certainly do so. An even more ambitious study would be to investigate how the elegance of the images depends on the entire multifractal spectrum of dimensions.

Chapter 3

Iterated Maps

This chapter will apply the methods in the previous chapters to the search for elegant attractors for a variety of iterated maps of a particular type. The main difference is in the form of the nonlinear map function. It will also describe the role of symmetry in the visual appeal of the resulting images.

3.1 Six-dimensional Cubic Maps

The previous chapter considered only a single example of an iterated map as given by Eq. (2.1) in which the nonlinearities are quadratic (the square of the variables). By searching the space of the twelve parameters (coefficients of the terms of the equation), many elegant examples of attractors were found and displayed. Quadratic functions are perhaps the algebraically simplest form of nonlinearity. However, there are countless other possibilities.

Consider, for example, a simple modification of Eq. (2.1) in which the quadratic terms are replaced by cubic terms:

Case B:

$$
\begin{aligned}
x_{n+1} &= a_1 x + a_2 x^3 + a_3 y + a_4 y^3 + a_5 z + a_6 z^3 \\
&\quad + a_7 u + a_8 u^3 + a_9 v + a_{10} v^3 + a_{11} w + a_{12} w^3 \\
y_{n+1} &= x \\
z_{n+1} &= y \\
u_{n+1} &= z \\
v_{n+1} &= u \\
w_{n+1} &= v.
\end{aligned} \tag{3.1}
$$

This map has a characteristic not shared by its quadratic kin. If all the variables are replaced by their negative, the equations remain unchanged

57

since only linear and cubic terms (terms with odd powers) are present. Such a system is said to be *inversion invariant*, and it leads to a special kind of symmetry. Either the solution has the same inversion invariance, or there is a symmetric pair of solutions that are accessed by changing the sign of all the initial conditions. The symmetry is present not only in the spatial variables (x, y, z) but also in the colors (u, v, w). This adds a degree of elegance to the solutions not shared by the previous examples.

Using the method in the previous chapter, a few hundred images were produced, two examples of which are shown in Figs. 3.1 and 3.2. Their identification code is prefaced with a B to denote a system in the form of Eq. (3.1). The images are unremarkable except for their inversion symmetry, which makes them appear slightly more ordered and thus arguably more elegant than the asymmetric ones in the previous chapter.

3.2 Skewness

The previous result suggests that some measure of symmetry might be useful for determining elegance. Unfortunately, there is no single such quantity because there are many kinds of symmetry [Stewart and Golubitsky (1992); Field and Golubitsky (1992); Gilmore and Letellier (2007)]. However, for the special case of inversion symmetry for a time-delay map such as Eq. (3.1), a useful quantity is the *skewness* defined by

$$SK = \lim_{N\to\infty} \frac{1}{N} \sum_{n=1}^{N} \left[\frac{x_n - \langle x \rangle}{\sigma} \right]^3 \tag{3.2}$$

where $\langle x \rangle$ is the average (or mean) value of x, and σ is the *standard deviation* given by

$$\sigma = \lim_{N\to\infty} \sqrt{ \frac{1}{N} \sum_{n=1}^{N} [x_n - \langle x \rangle]^2 }. \tag{3.3}$$

As a practical matter, a value of $N = 1 \times 10^7$ is usually sufficient to evaluate SK to three significant digits. Also since our interest is in the skewness of the visual pattern rather than the time series itself, it is better to take $\langle x \rangle = (x_{min} + x_{max})/2$ which corresponds to the center of the image, which in fact does correspond with $\langle x \rangle$ if the image is symmetric. Note that unlike the other measures defined in the previous chapter ($PF, CP, DF, \ldots$) that are always positive, the skewness can have either sign, with a positive

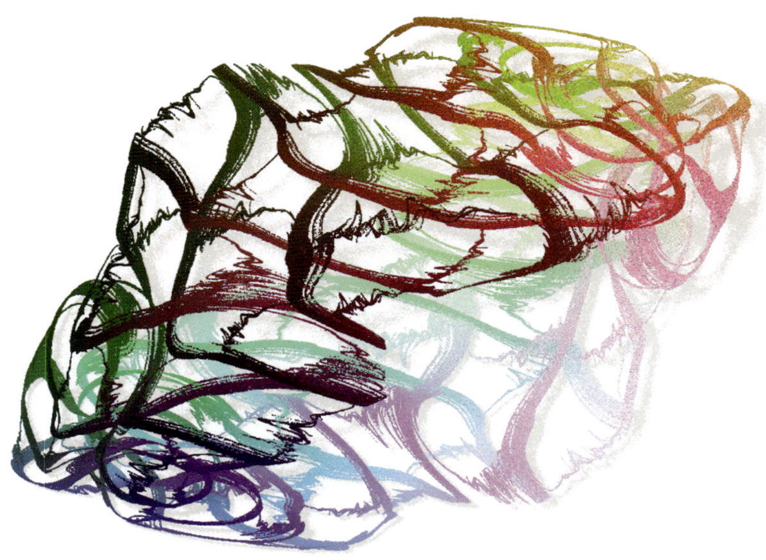

Fig. 3.1   `BRHPZELIRLTJM` ($PF = 0.383, CP = 0.673, LE = 0.0054, DF = 2.0266$).

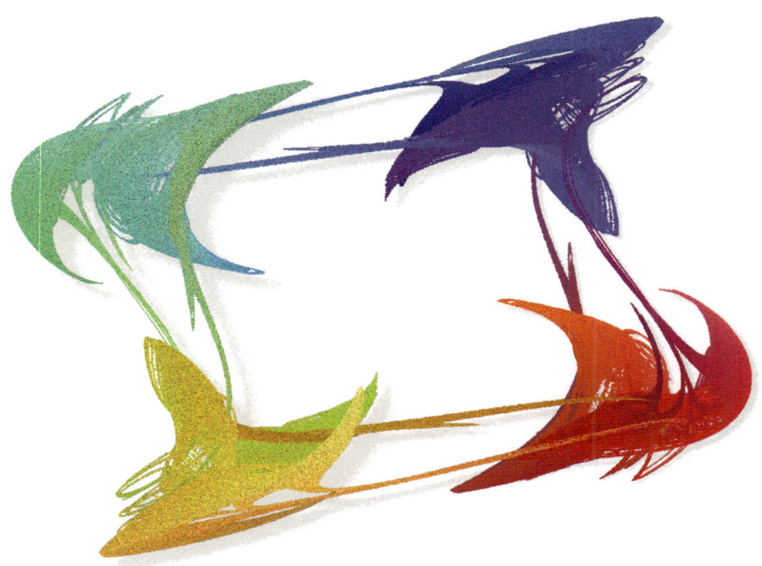

Fig. 3.2   `BLXFVRBRHILHY` ($PF = 0.256, CP = 0.857, LE = 0.0504, DF = 2.1025$).

value indicating that the attractor resides predominantly in the upper right and negative in the lower left.

The calculated skewness for Figs. 3.1 and 3.2 is $|SK| \approx 0.001$ rather than zero presumably because of the finite value of N and the fact that a chaotic orbit visits the different regions of the attractor in an irregular manner. However, the examples in the previous chapter have a decidedly non-zero skewness, with the largest being Fig. 2.4 with $SK = -0.606$ and the smallest being Fig. 2.7 with $SK = 0.014$. Furthermore, systems whose equations are inversion invariant such as Eq. (3.1) can have a symmetric pair of solutions that are not individually inversion invariant and hence have a non-zero skewness. Thus the value of SK allows you to identify coexisting solutions and eliminate (or select) them as desired.

Systems with multiple coexisting attractors are said to be *multistable*, and their existence can be problematic in engineering applications, although useful for some purposes. In any case, it is generally desirable to identify all the coexisting solutions of a system of nonlinear equations, and the only sure way to do that in general is to examine a large range of initial conditions using some criterion such as skewness to identify coexisting attractors since two attractors are not likely to have identical skewness.

3.3 Three-dimensional Cubic Maps

One problem with the previous six-dimensional maps is that their attractors tend to have a fractal dimension that is relatively too high to make them very elegant. In addition, the maps are missing cross terms such as $x^2 y$ and xy^2, which limits the variety of solutions. Both of these problems can be corrected by considering a three-dimensional scalar time-delay cubic map with all possible linear and cubic terms,

$$
\begin{aligned}
&\text{Case C:}\\
x_{n+1} &= (a_1 + a_2 x^2 + a_3 y^2 + a_4 z^2)x\\
&\quad + (a_5 + a_6 x^2 + a_7 y^2 + a_8 z^2)y\\
&\quad + (a_9 + a_{10} x^2 + a_{11} y^2 + a_{12} z^2)z\\
y_{n+1} &= x\\
z_{n+1} &= y\\
u_{n+1} &= z\\
v_{n+1} &= u\\
w_{n+1} &= v.
\end{aligned}
\tag{3.4}
$$

This map is inversion invariant with twelve parameters $(a_1, \ldots, a_{12})$ and six variables, but only three of those variables (x, y, z) contribute to the dynamics. The other three variables (u, v, w) control the colors as before, but they are slave variables that do not influence the dynamics. Consequently, any attractor for the system must have a dimension less than 3.0 and thus has a higher probability of being elegant.

Two solutions selected from a collection of a few dozen candidates are shown in Figs. 3.3 and 3.4. Their code is prefaced with a C to denote that they came from Eq. (3.4). The skewness is nearly zero because of the inversion invariance of the equations.

3.4   Other Odd and Even Functions

The functions x and x^3 are examples of *odd* functions as defined by the condition that $f(x) = -f(-x)$ for all values of x. Any other integer power $f(x) = x^n$ is also an odd function provided n is odd whether positive, $1, 3, 5, \ldots$, or negative, $-1, -3, -5, \ldots$. Conversely, an *even* function is defined by the condition that $g(x) = g(-x)$, for all x, the simplest examples of which are $g(x) = x^n$ with n even, $\ldots, -4, -2, 0, 2, 4, \ldots$. Of course the vast majority of functions are neither odd nor even, and they tend to produce attractors that are less elegant than those that have symmetry. For example, $e^x = 1 + x + x^2/2 + \ldots$ which has both even and odd terms.

Some common odd and even functions are the following:

Odd $f(x)$	Even $g(x)$
x^n, n odd	x^n, n even
$\operatorname{sgn} x$	$|x|$
$\sin x$	$\cos x$
$\sinh x$	$\cosh x$
$\arcsin x$	$\arccos x$.

From this table, you can construct additional, more complicated odd and even functions using the following rules:

- The sum or difference of two odd functions is odd.
- The sum or difference of two even functions is even.
- Any constant multiple of an odd function is odd.
- Any constant multiple of an even function is even.
- The product or quotient to two odd functions is even.
- The product or quotient of two even functions is even.

 Elegant Fractals

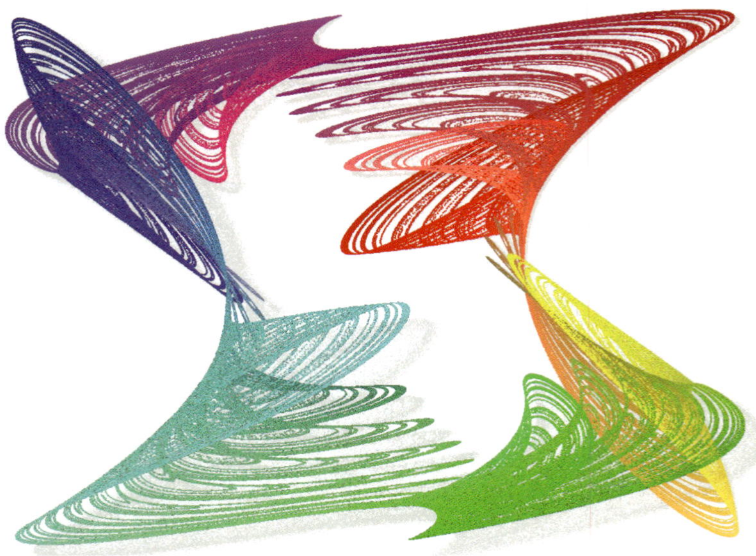

Fig. 3.3   CEQYJHBOXSQLL ($PF = 0.344, CP = 0.520, LE = 0.0928, DF = 1.7443$).

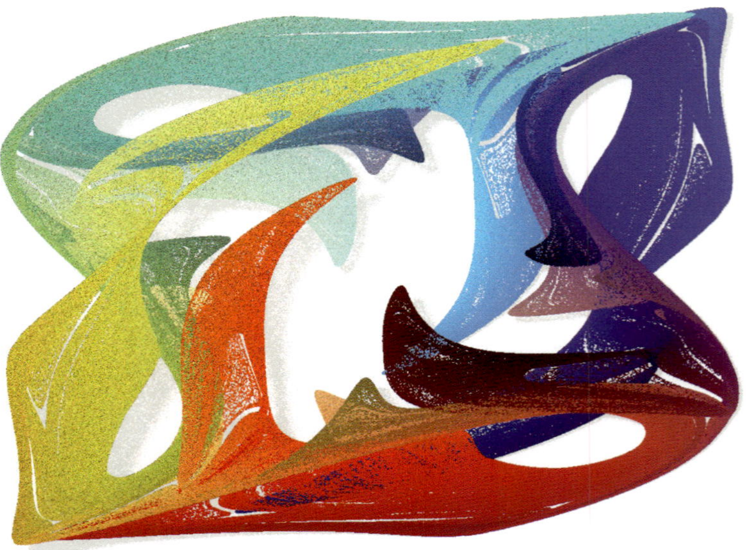

Fig. 3.4   CRYMVAYESWSOC ($PF = 0.622, CP = 0.801, LE = 0.1423, DF = 2.0442$).

- The product or quotient of an odd function and an even function is odd.
- The composition of two odd functions is odd.
- The composition of two even functions is even.
- The composition of an even function with an odd function is even.
- The composition of any function with an even function is even.

A *composition of functions* means that one function is the argument of the other, written as $(f \circ g)(x) = f(g(x))$. For example, if $f = \sin x$ and $g = x^2$ then $(f \circ g)(x) = \sin(x^2)$, which is even despite $\sin x$ being odd. The odd functions are particularly useful because they preserve the inversion symmetry of the equations and allow a wide variety of elegant fractal images.

3.5 Six-dimensional Signum Maps

Perhaps the simplest odd function that is not a power of x is the *signum function* $f(x) = \operatorname{sgn} x$, which is a function that is $+1$ for $x > 0$ and -1 for $x < 0$ and 0 for $x = 0$. The function is discontinuous (it jumps abruptly from one value to another when x passes through zero), but that does not pose any problem for our purposes.

Replacing the cubic terms in Eq. (3.1) with the signum function gives

Case **D**:

$$
\begin{aligned}
x_{n+1} &= a_1 x + a_2 \operatorname{sgn} x + a_3 y + a_4 \operatorname{sgn} y + a_5 z + a_6 \operatorname{sgn} z \\
&\quad + a_7 u + a_8 \operatorname{sgn} u + a_9 v + a_{10} \operatorname{sgn} v + a_{11} w + a_{12} \operatorname{sgn} w \\
y_{n+1} &= x \\
z_{n+1} &= y \\
u_{n+1} &= z \\
v_{n+1} &= u \\
w_{n+1} &= v.
\end{aligned}
\tag{3.5}
$$

Systems of this form are prefaced with a **D**, and two such examples are shown in Figs. 3.5 and 3.6. Because of the discontinuity of the signum function, the attractors tend to consist of numerous disconnected straight lines that intersect at sharp corners, although they can cluster densely, forming what looks like a surface as in Fig. 3.6.

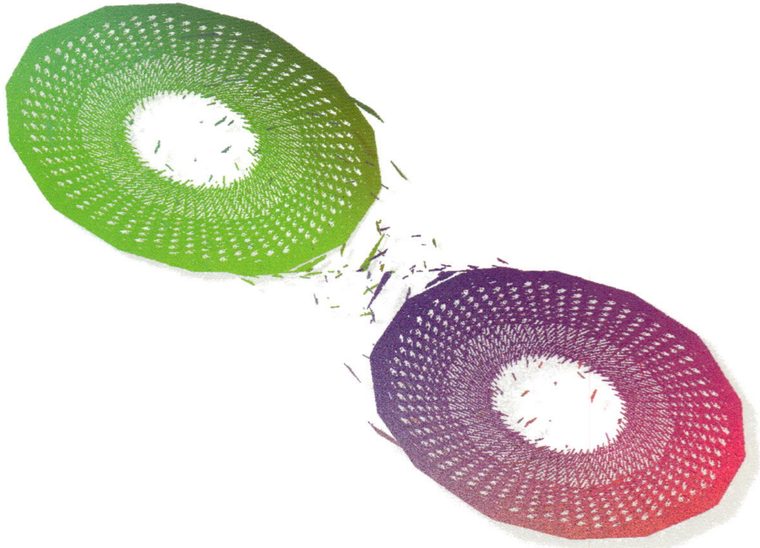

Fig. 3.5   **DFDNWRGVQMVIN** $(PF = 0.276, CP = 0.634, LE = 0.0010, DF = 2.0140)$.

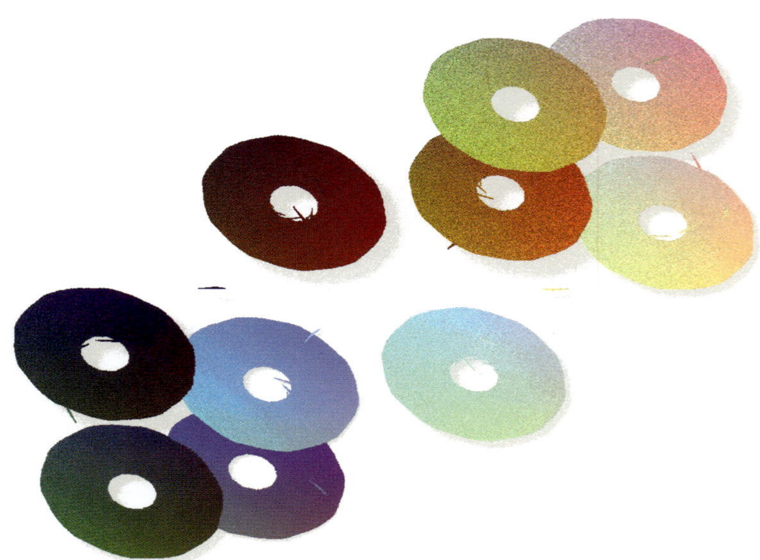

Fig. 3.6   **DORGMTHKKLEIS** $(PF = 0.374, CP = 0.979, LE = 0.0001, DF = 2.0019)$.

3.6 Four-dimensional Sine Maps

Another odd function is $f(x) = \sin x$, giving a four-dimensional system of the form

Case E:
$$
\begin{aligned}
x_{n+1} &= a_1 x + a_2 \sin(a_3 x) + a_4 y + a_5 \sin(a_6 y) \\
&\quad + a_7 z + a_8 \sin(a_9 z) + a_{10} u + a_{11} \sin(a_{12} u) \\
y_{n+1} &= x \\
z_{n+1} &= y \\
u_{n+1} &= z \\
v_{n+1} &= u \\
w_{n+1} &= v.
\end{aligned} \tag{3.6}
$$

The argument of the sine, as well as the other trigonometric functions in this chapter, is in radians, where 1 radian $= 180/\pi \approx 57.3$ degrees.

Two representative solutions are shown in Figs. 3.7 and 3.8 with codes prefaced by an E. They do not look much different from the previous cubic maps, which is not surprising since the Taylor series expansion of the sine is $\sin x = x - x^3/6 + \ldots$. (This is more properly called a *Maclaurin series* since it is evaluated in the vicinity of $x = 0$.)

3.7 Four-dimensional Tangent Maps

Other odd functions can be constructed by dividing an odd function such as the sine by an even function such as the cosine, which gives the *tangent function*

$$
\tan x = \frac{\sin x}{\cos x} = x + \frac{1}{3}x^3 + \ldots \tag{3.7}
$$

and a four-dimensional system of the form

Case F:
$$
\begin{aligned}
x_{n+1} &= a_1 x + a_2 \tan(a_3 x) + a_4 y + a_5 \tan(a_6 y) \\
&\quad + a_7 z + a_8 \tan(a_9 z) + a_{10} u + a_{11} \tan(a_{12} u) \\
y_{n+1} &= x \\
z_{n+1} &= y \\
u_{n+1} &= z \\
v_{n+1} &= u \\
w_{n+1} &= v.
\end{aligned} \tag{3.8}
$$

The Taylor series expansion for the tangent resembles that for the sine except that the signs of all the terms are positive rather than alternating,

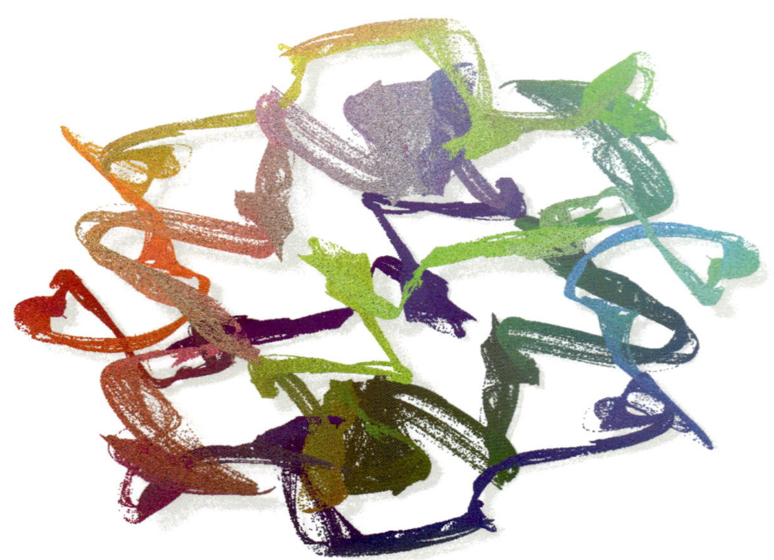

Fig. 3.7   **EIYKHMLFXFDCA** $(PF = 0.328, CP = 0.693, LE = 0.0373, DF = 2.0509)$.

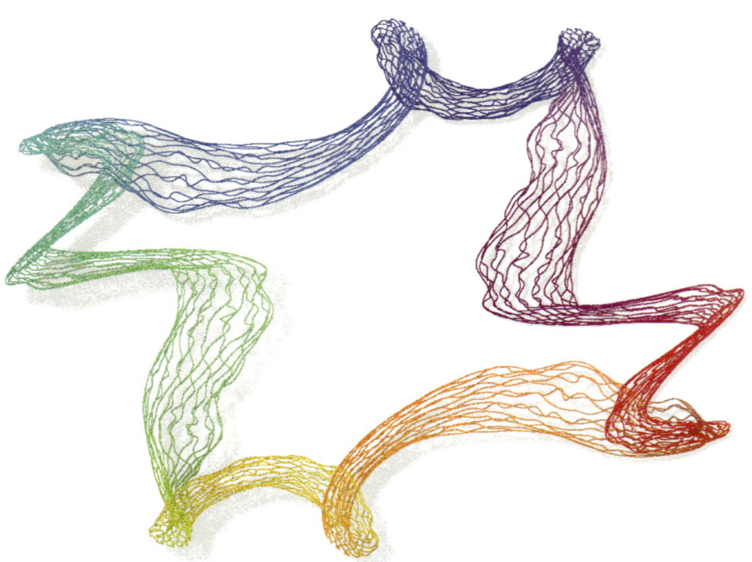

Fig. 3.8   **EVPIAIXQVYFMO** $(PF = 0.123, CP = 0.108, LE = 0, DF = 1)$.

$\tan x = x + x^3/3 + \ldots$. Unlike the sine function that is bounded in the range of -1 to $+1$, the tangent function becomes infinite at $x = \pm n\pi/2$, where n is an odd integer. This leads to more unbounded solutions but also to a wide variety of strange attractors, two examples of which are in Figs. 3.9 and 3.10, prefaced with an F to indicate that they came from Eq. (3.8).

3.8 Four-dimensional Hyperbolic Sine Maps

Another odd function that resembles the tangent is the *hyperbolic sine* given by

$$\sinh x = \frac{e^x - e^{-x}}{2} = x + \frac{1}{6}x^3 + \ldots \tag{3.9}$$

and giving a four-dimensional system of the form

Case G:
$$
\begin{aligned}
x_{n+1} &= a_1 x + a_2 \sinh(a_3 x) + a_4 y + a_5 \sinh(a_6 y) \\
&\quad + a_7 z + a_8 \sinh(a_9 z) + a_{10} u + a_{11} \sinh(a_{12} u) \\
y_{n+1} &= x \\
z_{n+1} &= y \\
u_{n+1} &= z \\
v_{n+1} &= u \\
w_{n+1} &= v.
\end{aligned} \tag{3.10}
$$

The Taylor series expansion for the hyperbolic sine resembles that for the tangent except that the hyperbolic sine approaches infinity only exponentially rather than explosively at $x = \pm\pi/2$. This leads to fewer unbounded solutions and a variety of strange attractors, two examples of which are in Figs. 3.11 and 3.12, prefaced with a G to indicate that they came from Eq. (3.10).

3.9 Four-dimensional Hyperbolic Tangent Maps

Another odd function that is the ratio of the hyperbolic sine to the hyperbolic cosine is the *hyperbolic tangent* given by

$$\tanh x = \frac{\sinh x}{\cosh x} = \frac{e^x - e^{-x}}{e^x + e^{-x}} = x - \frac{1}{3}x^3 + \ldots \tag{3.11}$$

 Elegant Fractals

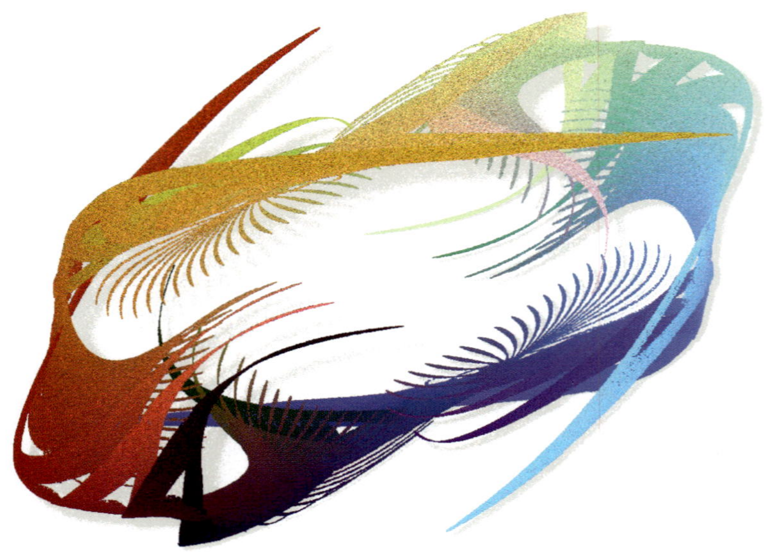

Fig. 3.9 **FMVTHNZLSDQMP** $(PF = 0.350, CP = 0.889, LE = 0.0766, DF = 2.1784)$.

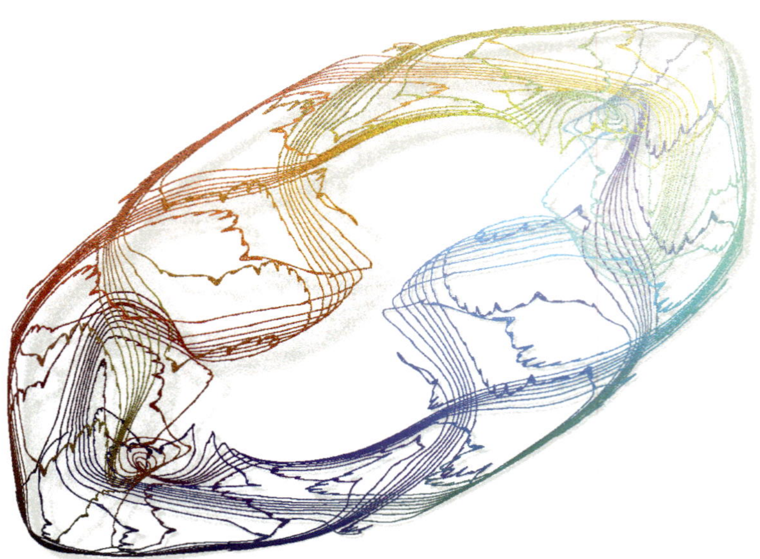

Fig. 3.10 **FQLTAUTAFAAID** $(PF = 0.164, CP = 0.181, LE = 0.0001, DF = 1.2875)$.

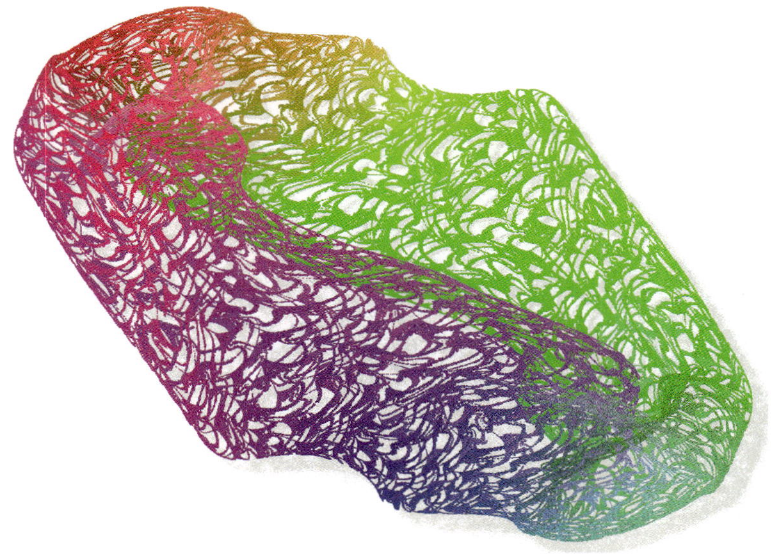

Fig. 3.11 **GILHTYLUDQATP** $(PF = 0.440, CP = 0.606, LE = 0.0005, DF = 1.5682)$.

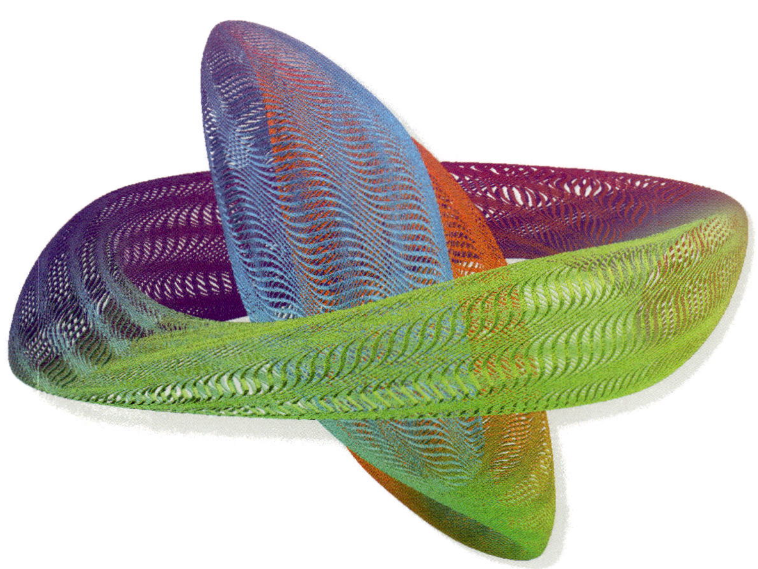

Fig. 3.12 **GLHKHGTSRFEAN** $(PF = 0.466, CP = 0.838, LE = 0, DF = 2)$.

and giving a four-dimensional system of the form

Case H:
$$\begin{aligned}
x_{n+1} &= a_1 x + a_2 \tanh(a_3 x) + a_4 y + a_5 \tanh(a_6 y) \\
&\quad + a_7 z + a_8 \tanh(a_9 z) + a_{10} u + a_{11} \tanh(a_{12} u) \\
y_{n+1} &= x \\
z_{n+1} &= y \\
u_{n+1} &= z \\
v_{n+1} &= u \\
w_{n+1} &= v.
\end{aligned} \tag{3.12}$$

Its Taylor series expansion resembles that for the sine, and it is similarly bounded in the range of -1 to $+1$ except that it does not oscillate as does the sine. Consequently, it is a more 'tame' function with fewer unbounded solutions, but it produces attractors that resemble the other previous cases, two examples of which are in Figs. 3.13 and 3.14, prefaced with an H to indicate that they came from Eq. (3.12).

The hyperbolic tangent is often used as the basis function in neural networks as Eq. (2.4) shows, and such networks can produce chaos and fractal patterns if they have feedback loops [Sprott (1998)]. In fact, Eq. (3.12) can be viewed as a *recurrent* neural network with four neurons.

3.10   Six-dimensional Integer Function Maps

The final example of an odd function that is of a type different from the previous cases involves rounding the variables to an integer denoted by $[x]$ and giving a six-dimensional system of the form

Case I:
$$\begin{aligned}
x_{n+1} &= a_1 x + a_2 [x] + a_3 y + a_4 [y] + a_5 z + a_6 [z] \\
&\quad + a_7 u + a_8 [u] + a_9 v + a_{10} [v] + a_{11} w + a_{12} [w] \\
y_{n+1} &= x \\
z_{n+1} &= y \\
u_{n+1} &= z \\
v_{n+1} &= u \\
w_{n+1} &= v.
\end{aligned} \tag{3.13}$$

This system produces relatively few elegant attractors because most cases have a ragged and colorless appearance. However, two interesting and rather different cases are shown in Figs. 3.15 and 3.16, prefaced with an I to indicate that they came from Eq. (3.13).

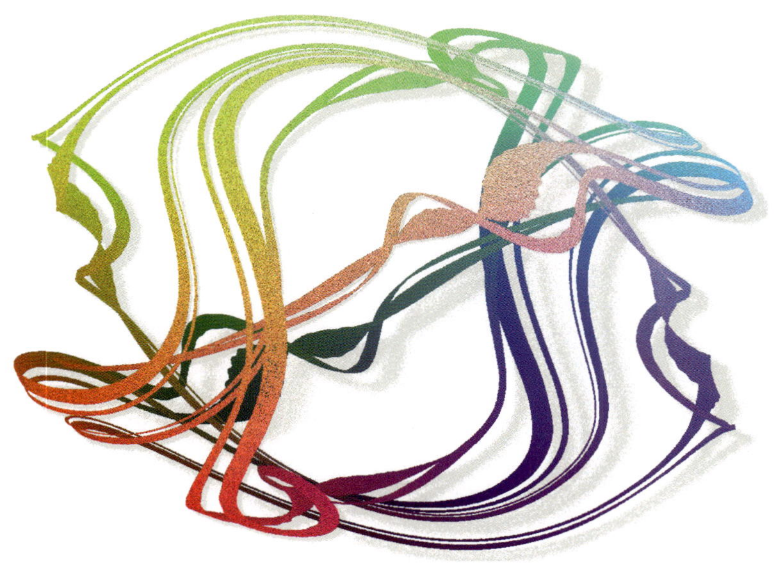

Fig. 3.13 **HDWUQTAEHPPFW** ($PF = 0.265, CP = 0.861, LE = 0.0015, DF = 2.0373$).

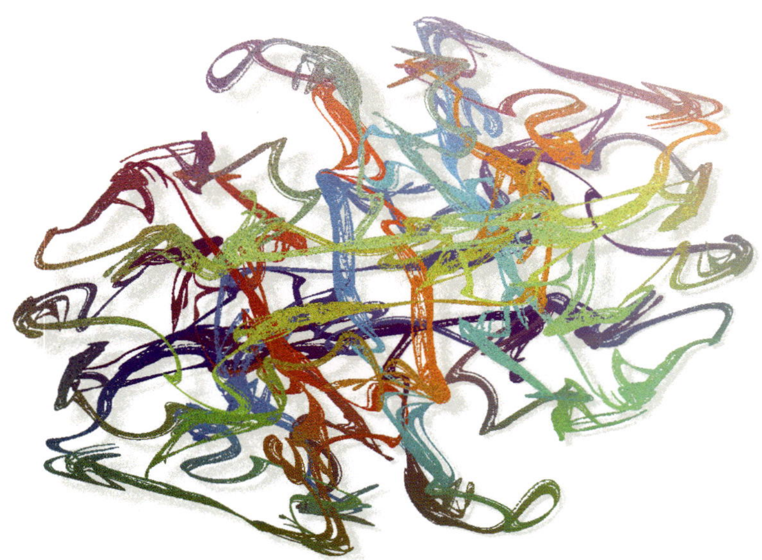

Fig. 3.14 **HLYTARVWZADMP** ($PF = 0.302, CP = 0.680, LE = 0.0138, DF = 1.8797$).

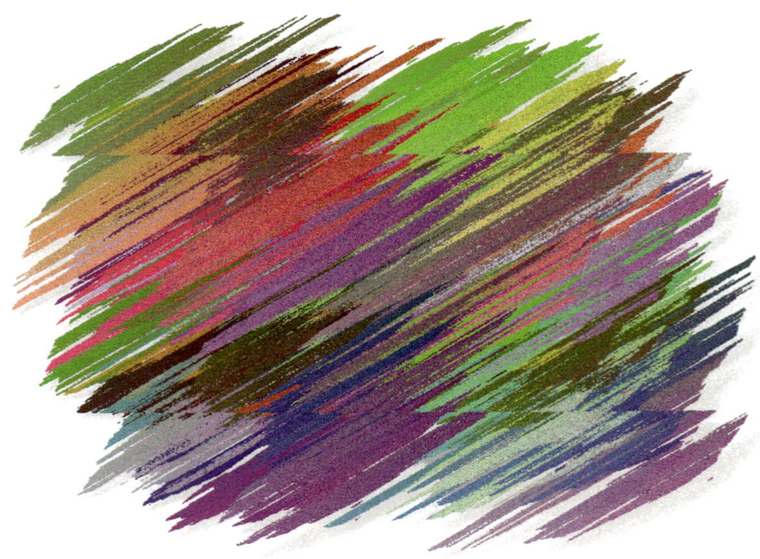

Fig. 3.15   `IHLXAVESHNMKM` ($PF = 0.654, CP = 0.904, LE = 0.2203, DF = 3.0303$).

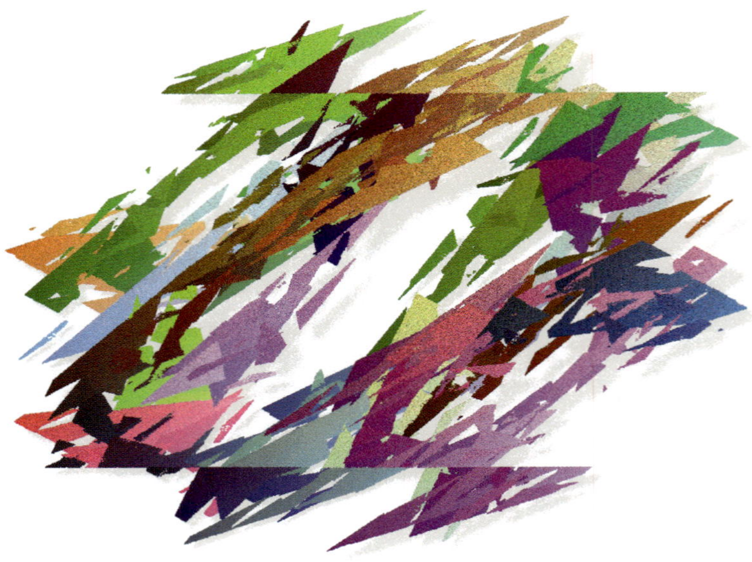

Fig. 3.16   `IQKPLCUNNKRKO` ($PF = 0.459, CP = 0.945, LE = 0.0185, DF = 3.0866$).

3.11 Symmetric Icons

Inversion symmetry in which the equations remain unchanged if the sign of all the variables are changed is only one of many kinds of symmetry. If the equations are invariant when only the sign of x and y is changed, the result is a system that is *rotationally symmetric* under a 180-degree rotation about the z-axis and an attractor that usually has the same symmetry, although a symmetric pair of attractors is also possible. However, scalar time-delay maps such as those in this chapter do not allow such a symmetry for these cases.

If a system is invariant when the sign of a single variable is changed, it has *reflection symmetry* with respect to a plane at the zero value of that variable. The scalar time-delay maps in this chapter do not allow those cases either. In fact, no combination of signs can be changed in these systems other than all of them, and so inversion invariance is the only possible such symmetry.

You can construct more complicated maps that have particular desired symmetries [Stewart and Golubitsky (1992); Field and Golubitsky (1992)], but a simpler method is to construct the attractors as we have been doing, and then introduce the symmetry through some transformation of variables before plotting the solution [Sprott (1996)]. The resulting technique is both simple and general.

Suppose, for example that you want to make a fractal that is symmetric with respect to a rotation through some angle about the z-axis like the petals of a flower. The angle should be such that an integer number N of rotations fits within 360 degrees (2π radians) so that each segment has an angular width of $2\pi/N$ radians. The task then is to distort the attractor so that it fits into a pie-shaped wedge and then plot it N times rotated through the appropriate angle.

If x_p an y_p are the horizontal and vertical pixel coordinates ($0 \leq x_p < w, 0 \leq y_p < h$) calculated for plotting the attractor in the conventional way, you can calculate a radius and angle using

$$r = \frac{x_p}{w}$$

$$\theta = \frac{2\pi(y_p/h + i)}{N}$$

(3.14)

where w is the number of horizontal pixels (1200 for these examples), h is the number of vertical pixels (900 for these examples), and i is an integer chosen randomly in the range of 0 to $N - 1$ that controls the segment in which each point is plotted. Typical values of N are 1 to 9. Choosing the segment randomly instead of just replicating each point N times gives the segments minute but aesthetically desirable small differences, helping to disguise their machine-generated origin.

Then plot each point at the position

$$x_p = \frac{(1 - r\sin\theta)w}{2}$$
$$y_p = \frac{(1 + r\cos\theta)h}{2}.$$

$$(3.15)$$

The sine and cosine are chosen in a way to enhance the right/left symmetry when N is odd, which is more common in nature (consider an upright human figure) than up/down symmetry, and hence is more visually appealing.

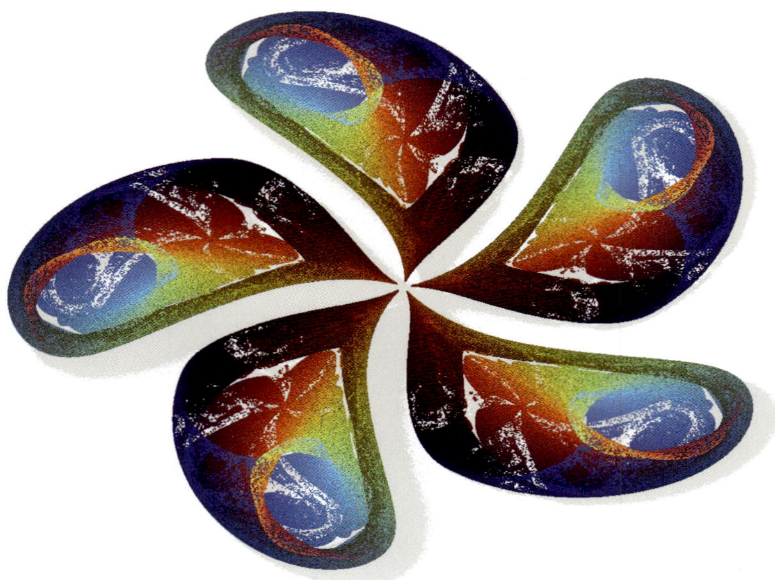

Fig. 3.17   CJLFWYOFEVTPC5 ($PF = 0.444, CP = 0.838, LE = 0.0639, DF = 2.0790$).

The signs are chosen so that the case with $N = 1$ most nearly resembles the untransformed case.

One additional embellishment is to reverse the angle of alternate segments whenever N and i are even integers by replacing θ with $2\pi/N - \theta$, which adds another form of reflection symmetry about certain radials.

Images formed in this way have been called *symmetric icons* [Stewart and Golubitsky (1992); Sprott (1996)], some examples of which are in Figs. 3.17 to 3.23. The code reflects the system from which they were generated, with a single digit appended to the end to indicate the number of segments (N).

Is there an optimal number of segments that maximizes the elegance? Do you prefer ones with an odd or an even number of segments? Do you like them better than or not as much as the previous examples? Do you think their resemblance to flowers enhances their elegance?

Perhaps the images in this chapter remind you of paintings by the German artist Paul Klee (1998–2004).

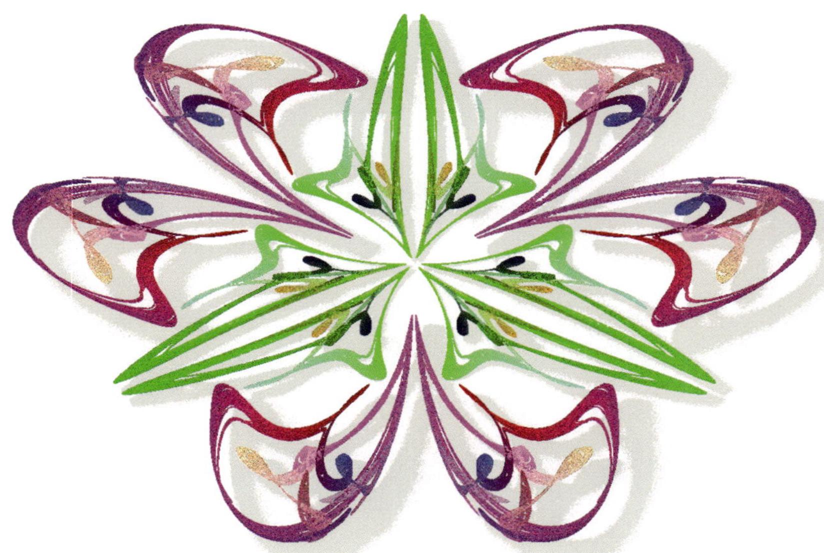

Fig. 3.18   **CFWJTWIIDEPJR6** ($PF = 0.213, CP = 0.707, LE = 0.1365, DF = 1.7891$).

     Elegant Fractals

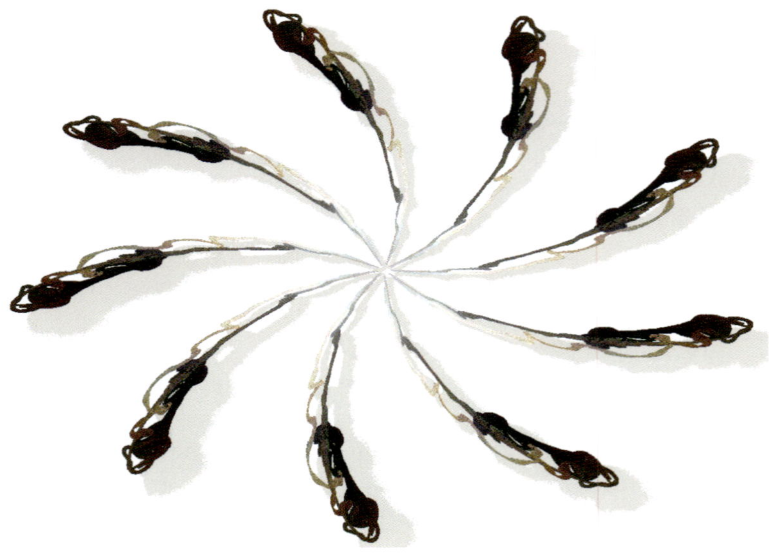

Fig. 3.19   ESRQRNJRTOGAY9 ($PF = 0.100, CP = 0.737, LE = 0.0454, DF = 2.1410$).

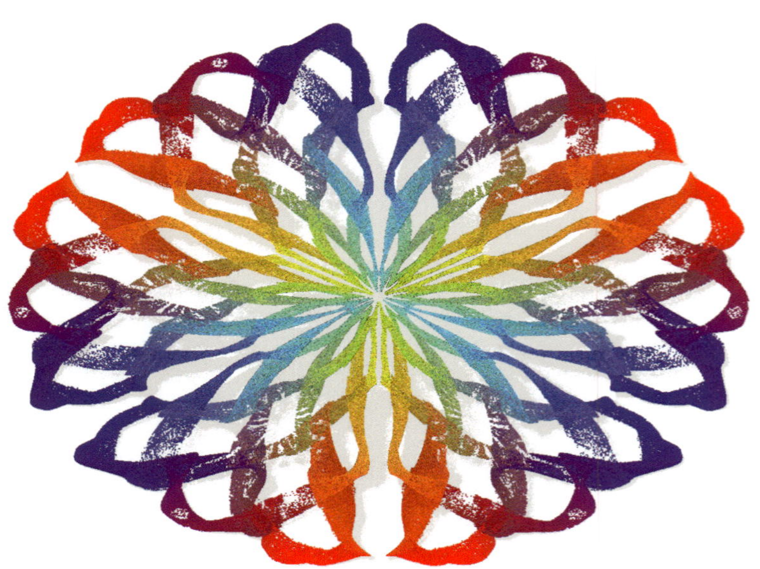

Fig. 3.20   EVLYBSKVTVKYC6 ($PF = 0.409, CP = 0.685, LE = 0.0426, DF = 2.1451$).

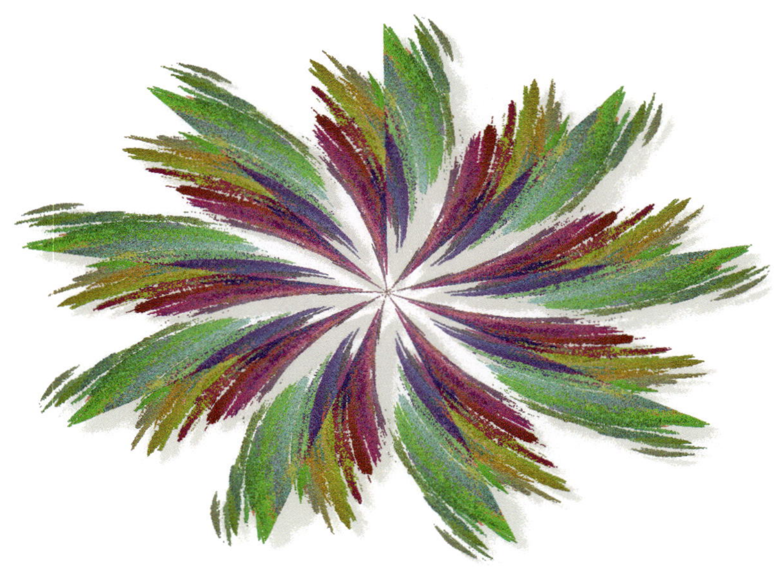

Fig. 3.21 `DTGOJJRUANPKP9` ($PF = 0.404, CP = 0.803, LE = 0.1310, DF = 2.8739$).

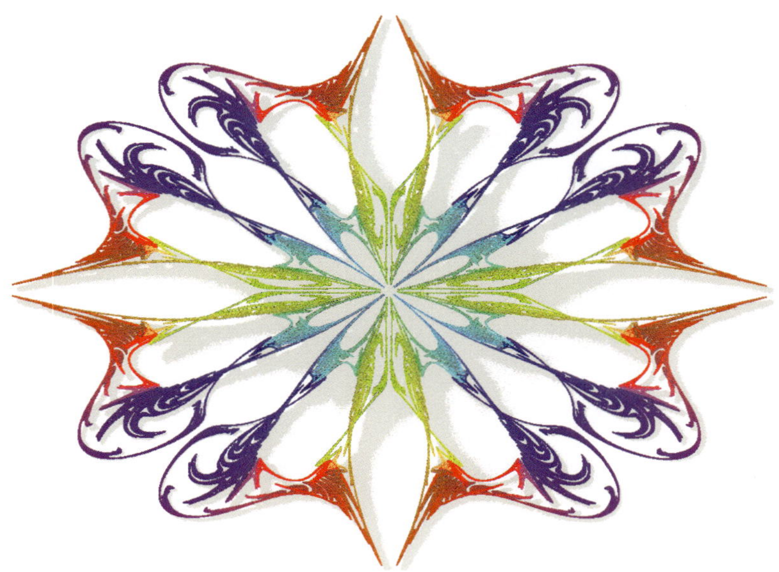

Fig. 3.22 `CPENDBQIYPMDN8` ($PF = 0.197, CP = 0.603, LE = 0.0488, DF = 1.8000$).

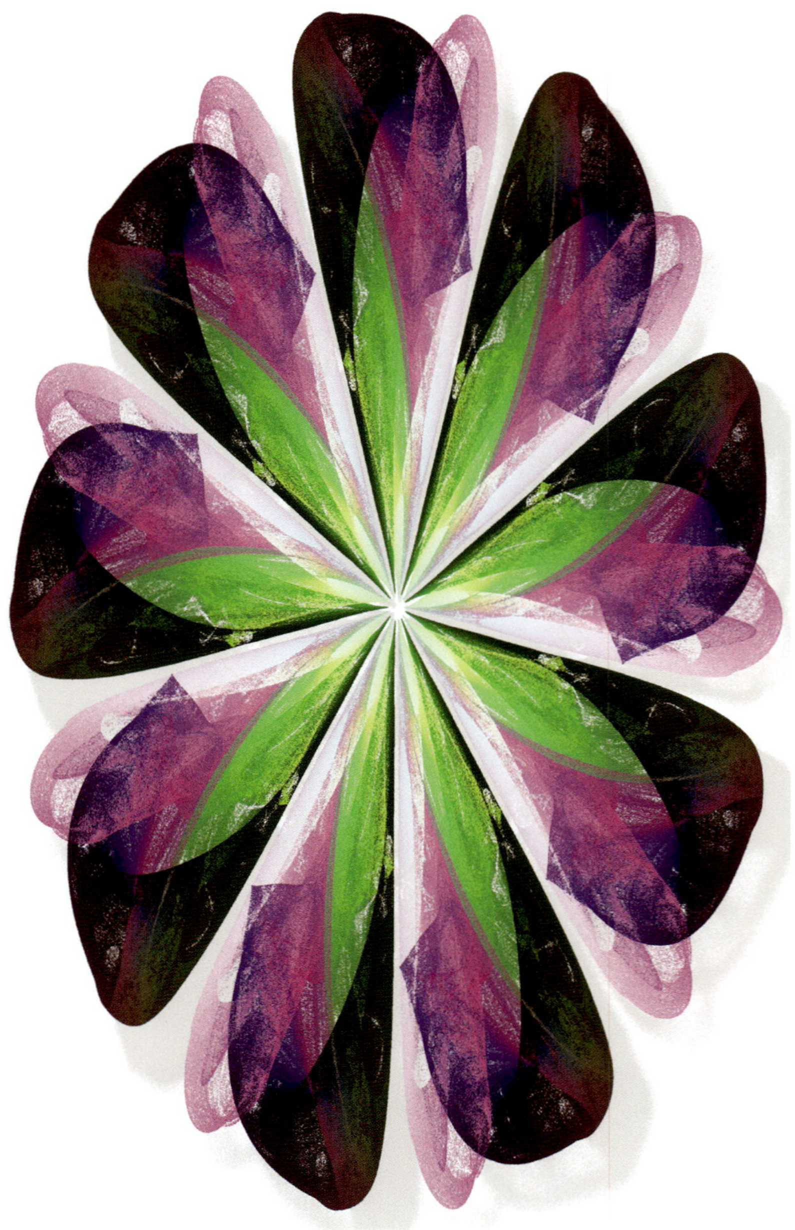

Fig. 3.23   GONRUMLWIYNKQ9 $(PF = 0.625, CP = 0.660, LE = 0.0826, DF = 3.1085)$.

Chapter 4

Chaotic Flows

Most physical processes in nature are better described by variables that change continuously in time rather than at discrete instants. Such systems are governed by systems of ordinary differential equations rather than by iterated maps. This chapter give examples of continuous-time systems whose attractors have a somewhat different appearance than those in the preceding chapters.

4.1　Three-dimensional Quadratic Flows

Systems in which the variables change continuously in time are more realistic but are more difficult to solve and have a lesser chance of exhibiting chaos. In particular, the *Poincaré–Bendixson theorem* [Hirsch *et al.* (2004)] says that chaos can only exist if there are at least three variables. The reason is simply that two trajectories must not intersect because that would imply that the point of intersection has two different futures, which is inconsistent with a deterministic dynamical system. Thus, unlike iterated maps, where the iterates can jump over one another even in one or two dimensions, a third dimension is required to allow continuous trajectories to pass beneath or above one another and wrap around into a strange attractor.

Such systems are called 'flows' because you can imagine each point in space represented by a velocity vector indicating the direction and speed with which an object located at that point would move. The collection of all such points then moves something like a flowing and often turbulent (or chaotic) fluid. Thus if these images remind you of swirling liquids or weather patterns, that is no accident.

As with iterated maps, a nonlinearity is required to produce chaos, and the simplest such nonlinearity is quadratic. Thus we begin with a relatively simple, three-dimensional quadratic system of the form

Case J:
$$\dot{x} = (a_1 + a_2 x + a_3 y + a_4 z)y$$
$$\dot{y} = (a_5 + a_6 x + a_7 y + a_8 z)z$$
$$\dot{z} = (a_9 + a_{10} x + a_{11} y + a_{12} z)x \qquad (4.1)$$
$$u = \dot{x}$$
$$v = \dot{y}$$
$$w = \dot{z}.$$

This system has been constructed to have twelve parameters as with the previous iterated maps since that is a number sufficient to give an essentially unlimited variety of cases while not being excessively complicated. It also has a kind of *circulant symmetry* since $\dot{x}$ is proportional to y, $\dot{y}$ is proportional to z, and $\dot{z}$ is proportional to x. Symmetries generally produce more elegant solutions, although in this case, the symmetry is broken by the fact that the coefficients in the first three equations are generally different.

In Eq. (4.1), the dot over a symbol denotes the time derivative ($\dot{x} \equiv dx/dt$), a terminology used by Isaac Newton, while the more common and more unwieldy (and less elegant) form was used by his rival Gottfried Leibniz [Cajori (1993)]. The first three equations in Eq. (4.1) constitute a three-dimensional system of *ordinary differential equations*, while the last three equations without dots (for u, v, and w) are algebraic equations that do not participate in the dynamics but are slave variables that allow control of the three colors. They are the time derivatives of the three respective dynamical variables.

Now we are confronted with how to solve such a system of ordinary differential equations on a digital computer. Books have been written on that topic [Milne (1970); Gear (1971); Atkinson *et al.* (2009); Griffiths and Higham (2010)], and it is still an area of active research. However, since our interest is in making elegant fractals rather than providing accurate solutions, it suffices to use the *Euler method*, which is particularly simple, although all the cases in this chapter have been checked using the far more accurate fourth-order *Runge–Kutta method* with an adaptive step size and error control [Press *et al.* (2007)]. The Euler method applied to Eq. (4.1) reduces it to an iterated map,

$$
\begin{aligned}
u &= (a_1 + a_2 x + a_3 y + a_4 z)y \\
v &= (a_5 + a_6 x + a_7 y + a_8 z)z \\
w &= (a_9 + a_{10} x + a_{11} y + a_{12} z)x \\
x_{n+1} &= x + hu \\
y_{n+1} &= y + hv \\
z_{n+1} &= z + hw.
\end{aligned}
\tag{4.2}
$$

The new constant h is the *iteration step size*, and it will be taken as 0.001, which is sufficiently small to give a good approximation to the solution of the differential equations but large enough to allow efficient computation. The ordering of the equations has been reversed to emphasize that u, v, and w must be evaluated first before x, y, and z are iterated. If you evaluate x, y, and z first, be careful to update the three variables simultaneously rather than sequentially, which will require storing the updated values as temporary variables. Evaluating them sequentially is called the *leapfrog method*, and it has some advantages, but it will give a slightly different result.

Thereafter, the solution follows as if it is just another kind of iterated map in which each iterate is very close to its predecessor. In fact the iterates are so close that it is generally unnecessary to connect the dots with lines in plotting the solution, which is in the form of one long, continuous curve. You can use all the selection methods described in the previous chapter to eliminate cases that are not likely to give elegant fractals.

Even with $h = 0.001$, ten million iterates allows the solution to be followed for ten thousand time units, which is usually more than enough to capture the nature of the attractor. In fact, for a chaotic flow, the fractal dimension of the attractor will be greater than 2.0, and so it is best not to run the calculation for too long lest the plot fill in so densely that you cannot observe the individual orbits.

Figures 4.1 to 4.4 show a variety of cases plotted in the same way as for the iterated maps in the previous chapter with their codes prefaced with a J to indicate that they were produced by Eq. (4.1). The values of LE and DF shown in the captions were determined using a more accurate integrator run for a much longer time and typically agree with the values obtained from the Euler integrator within about a factor of two, although the plots of the attractors usually are indistinguishable for the two integrators.

Note that the Kaplan–Yorke dimension DF is only slightly greater than 2.0 for all the cases, and this is typical for three-dimensional dissipative

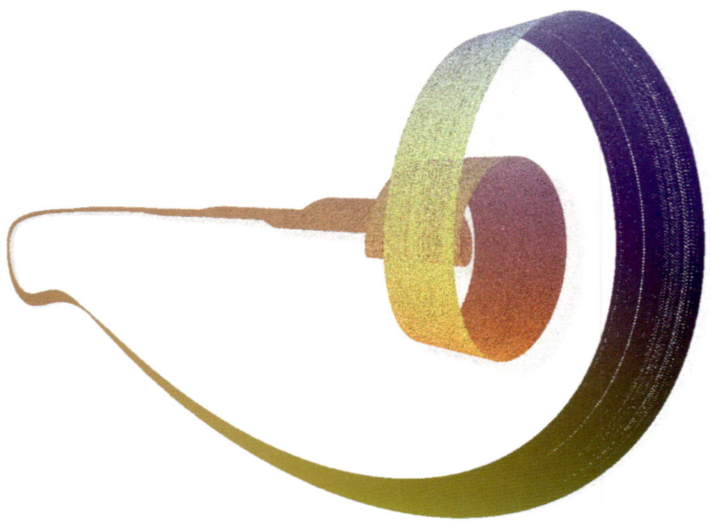

Fig. 4.1   JFCNKPVOADRVG $(PF = 0.224, CP = 0.651, LE = 0.0351, DF = 2.0247)$.

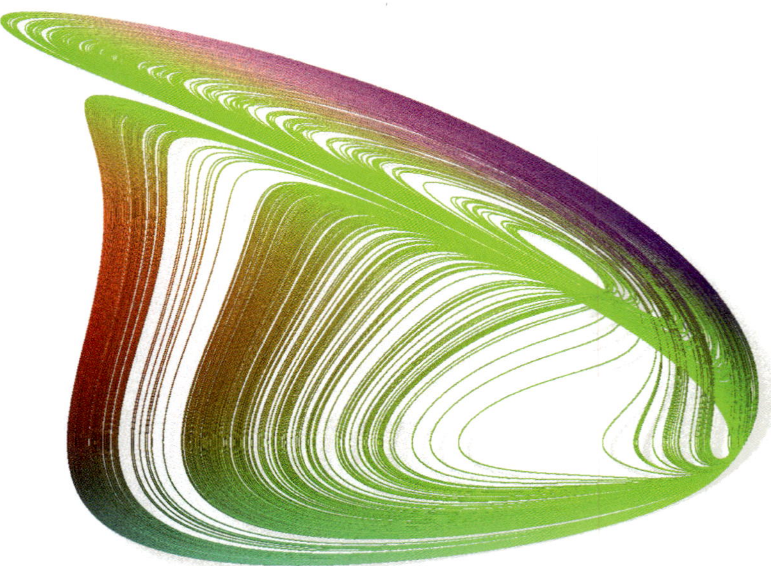

Fig. 4.2   JRIIHZONSNXSX $(PF = 0.394, CP = 0.649, LE = 0.0418, DF = 2.0546)$.

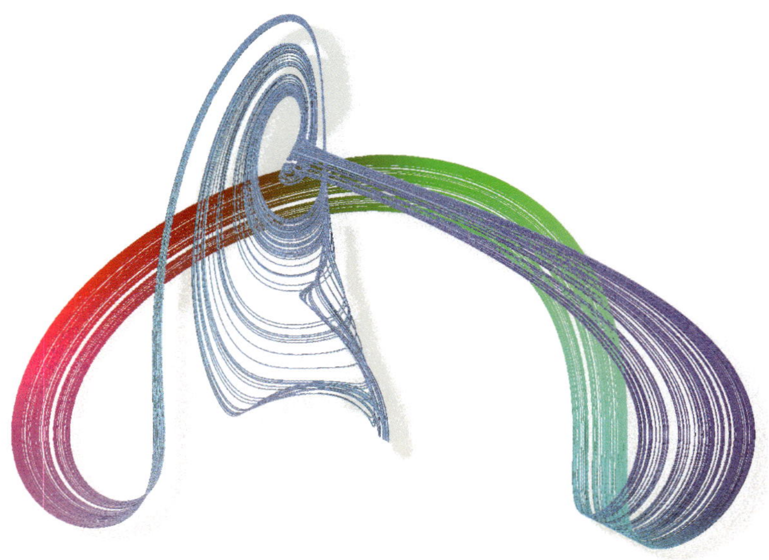

Fig. 4.3 **JXHAPQZMPOZUY** $(PF = 0.224, CP = 0.489, LE = 0.0082, DF = 2.0063)$.

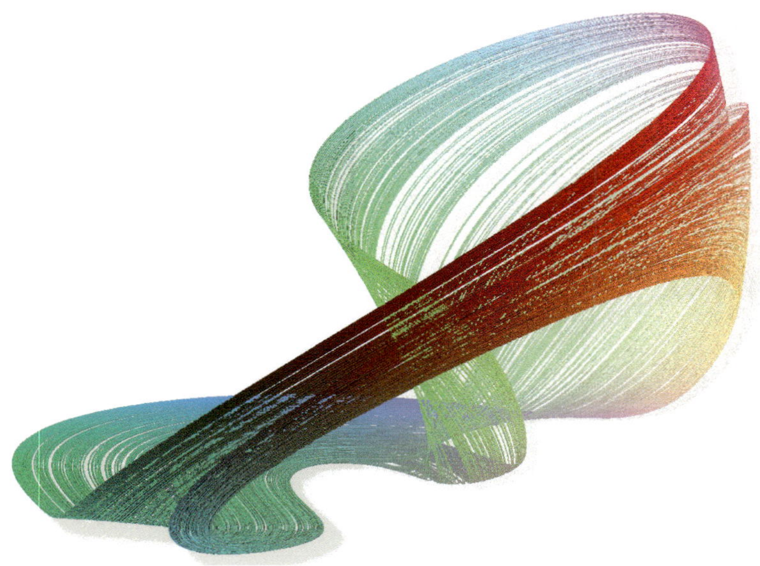

Fig. 4.4 **JXKDHSAOMJDUO** $(PF = 0.389, CP = 0.679, LE = 0.0323, DF = 2.0626)$.

chaotic systems. Because the dimension is not an integer, the resulting attractor is necessarily a fractal, and zooming in on a portion of it will reveal that the long continuous line is entangled with itself on arbitrarily small scales while never intersecting itself.

Also note that the Lyapunov exponents are per unit time rather than per iteration as in the previous chapter and that the latter is smaller than the former by a factor of $h = 0.001$, reflecting the fact that many (1000) iterations of the map are required to advance the flow by one time unit.

Of course, any of these cases can be plotted as symmetric icons, and Figs. 4.5 and 4.6 show two such examples.

4.2   Three-dimensional Cubic Jerk Flows

All the discrete-time systems in the previous chapter have counterparts in continuous-time flows, and most of the considerations discussed there carry over to the cases in this chapter. For example, the systems can be made inversion symmetric by limiting the nonlinearities to odd functions, the simplest example of which is the cubic. Thus Eq. (3.4) has a continuous-time counterpart,

$$
\begin{aligned}
&\text{Case K:}\\
&\dot{x} = y\\
&\dot{y} = z\\
&\dot{z} = (a_1 + a_2 x^2 + a_3 y^2 + a_4 z^2)x\\
&\qquad\; + (a_5 + a_6 x^2 + a_7 y^2 + a_8 z^2)y\\
&\qquad\; + (a_9 + a_{10} x^2 + a_{11} y^2 + a_{12} z^2)z\\
&u = \dot{x}\\
&v = \dot{y}\\
&w = \dot{z}.
\end{aligned}
\tag{4.3}
$$

The ordering of x, y, and z has been reversed from Eq. (3.4) according to convention.

Equation (4.3) is an example of a *jerk system*. The reason for the peculiar name is that if you consider x as the position of a moving object, then $\dot{x} = y$ is its velocity, $\ddot{x} = \dot{y} = z$ is its acceleration, and $\dddot{x} = \ddot{y} = \dot{z}$ is the rate of change of its acceleration, which is called a 'jerk' [Schot (1978)]. The right-hand side of the $\dot{z}$ equation is the *jerk function*, and it can be written in scalar form as $\dddot{x} = (a_1 + a_2 x^2 + a_3 \dot{x}^2 + a_4 \ddot{x}^2)x + (a_5 + a_6 x^2 + a_7 \dot{x}^2 + a_8 \ddot{x}^2)\dot{x} + (a_9 + a_{10} x^2 + a_{11} \dot{x}^2 + a_{12} \ddot{x}^2)\ddot{x}$.

Any three-dimensional dynamical system with a single nonlinearity, as well as many others, can be transformed into jerk form [Eichhorn *et al.*

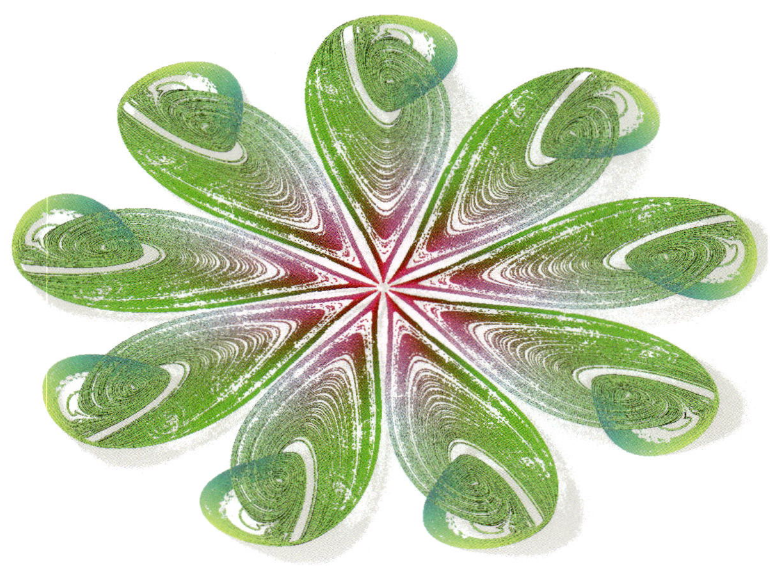

Fig. 4.5   JRHJHGVVBHIPZ9 $(PF = 0.445, CP = 0.559, LE = 0.0233, DF = 2.0196)$.

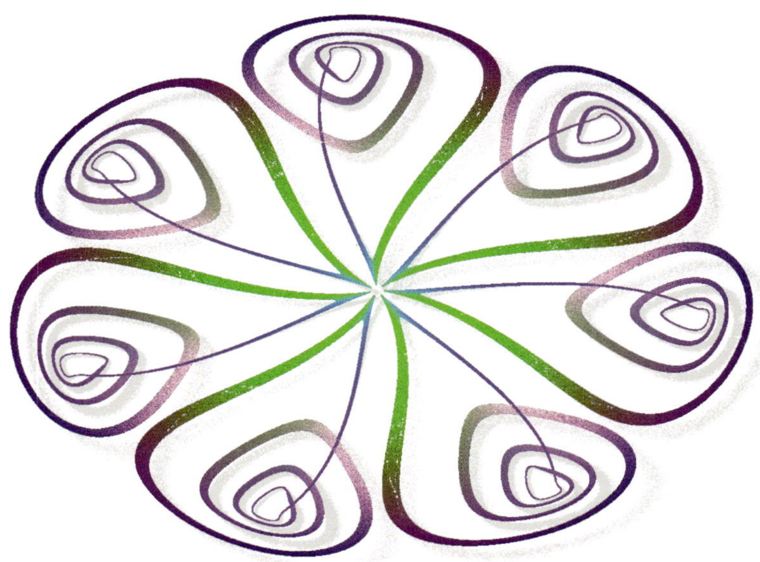

Fig. 4.6   JUYJBVTKOPNQR7 $(PF = 0.179, CP = 0.658, LE = 0.0478, DF = 2.0215)$.

(1998, 2002)], and so such a system is rather more general than it appears. Furthermore, since chaos in flows requires as least three dimensions, chaotic mechanical motion implies that there must be jerks (changing accelerations) in the equations of motion. Jerk systems are the continuous-time analog of the discrete-time, time-delay maps in the previous chapters.

Some examples of attractors from cubic jerk flows are shown in Figs. 4.7 and 4.8 prefaced with a K to indicate that they were produced by Eq. (4.3).

4.3　Four-dimensional Sine Hyperjerk Flows

The idea of a jerk system can be extended to higher dimensions, in which case it is called a *hyperjerk* [Chlouverakis and Sprott (2005)]. The time derivative of a jerk (or the fourth derivative of position) has been called a 'snap' with successively higher derivatives 'crackle' and 'pop' [Sprott (1997b, 2010)]. Such an example in four dimensions using the sine nonlinearity can be constructed by analogy with Eq. (3.6),

$$
\begin{aligned}
&\text{Case L:}\\
&\dot{x} = y\\
&\dot{y} = z\\
&\dot{z} = u\\
&\dot{u} = a_1 x + a_2 \sin(a_3 x) + a_4 y + a_5 \sin(a_6 y)\\
&\qquad + a_7 z + a_8 \sin(a_9 z) + a_{10} u + a_{11} \sin(a_{12} u)\\
&v = \dot{x}\\
&w = \dot{y}.
\end{aligned}
\tag{4.4}
$$

The sine nonlinearity is better behaved than the square since it is bounded between -1 and $+1$, thus reducing the prevalence of unbounded solutions, although certainly not eliminating them.

Two examples of such a system are given in Figs. 4.9 and 4.10 with codes prefaced by L to indicate that they came from Eq. (4.4).

4.4　Four-dimensional Hyperbolic Tangent Hyperjerk Flows

The cases in Eqs. (3.7) and (3.8) that involve the tangent and hyperbolic sine functions do not lend themselves to a corresponding flow because they go to infinity at a finite value of their argument, and as a result, the Euler method is not adequate for their calculation. Furthermore, the resulting attractors are not greatly different from other systems of ordinary differential equations that are better behaved.

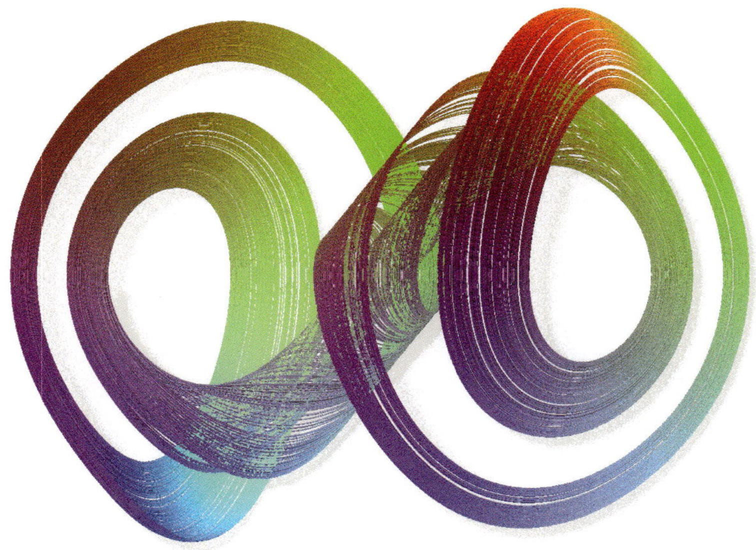

Fig. 4.7 **KXHAVGMFSPECJ** ($PF = 0.471, CP = 0.802, LE = 0.0713, DF = 2.0312$).

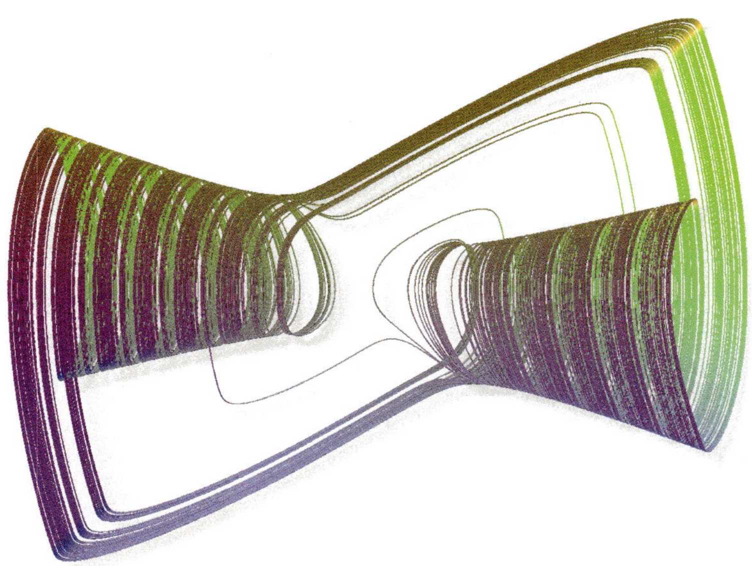

Fig. 4.8 **KIMOHOCWAWWCF** ($PF = 0.317, CP = 0.619, LE = 0.0781, DF = 2.0018$).

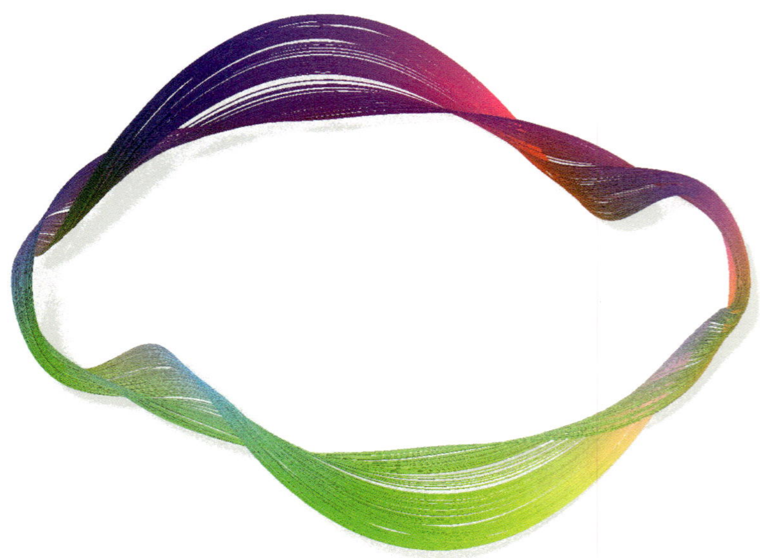

Fig. 4.9 **LLDALQSAQAIJI** ($PF = 0.201, CP = 0.774, LE = 0.0330, DF = 2.2146$).

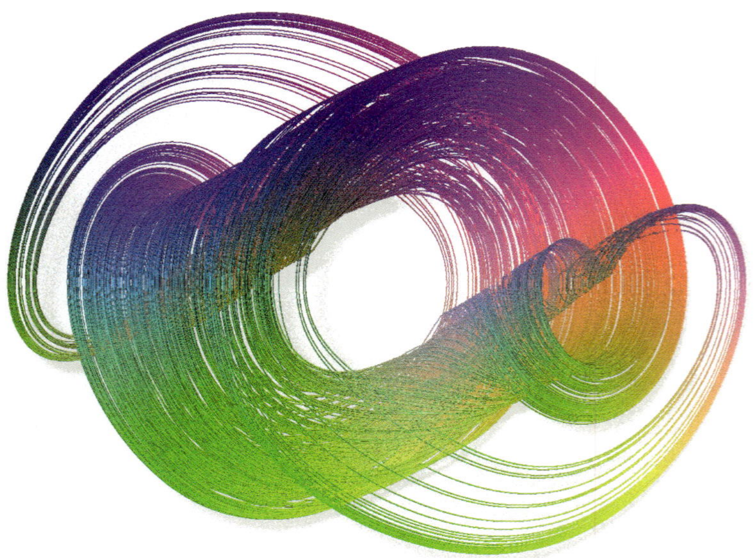

Fig. 4.10 **LLXFGYOCEKEOD** ($PF = 0.489, CP = 0.626, LE = 0.1105, DF = 2.3085$).

By contrast, the hyperbolic tangent function in Eq. (3.9) is well behaved because it is bounded in the range of -1 to $+1$, giving the system,

Case M:
$$\dot{x} = y$$
$$\dot{y} = z$$
$$\dot{z} = u$$
$$\dot{u} = a_1 x + a_2 \tanh(a_3 x) + a_4 y + a_5 \tanh(a_6 y)$$
$$\qquad + a_7 z + a_8 \tanh(a_9 z) + a_{10} u + a_{11} \tanh(a_{12} u)$$
$$v = \dot{x}$$
$$w = \dot{y}.$$
$$(4.5)$$

However, the weaker nonlinearity makes chaotic solutions somewhat less common.

Two examples of such a system are given in Figs. 4.11 and 4.12 with codes prefaced by M to indicate that they came from Eq. (4.5). Figure 4.11 is an attracting torus that will eventually fill in the entire two-dimensional surface to the resolution of the plot after a time of about 3.3×10^4 or 3.3×10^7 iterations.

4.5 Generalized Lorenz System

The previous hyperjerk systems produce a variety of strange attractors, but there is a certain sameness about them, which is due in part to the fact that their nonlinearities are odd functions reasonably approximated by a superposition of linear and cubic terms. However, there are many well-known chaotic systems that are not in the form of any of the previous systems. One such example is the famous *Lorenz system* [Lorenz (1963)],

$$\dot{x} = \sigma(y - x), \quad \dot{y} = -xz + rx - y, \quad \dot{z} = xy - bz, \qquad (4.6)$$

which has chaotic solutions for $(\sigma, r, b) = (10, 28, 8/3)$. This was the first chaotic system that was shown to have a strange attractor, and it is still often used as the prototypical example of a chaotic flow.

Adding a few extra terms to Eq. (4.6), bringing the total to 12 as usual, gives a more general version that includes the Lorenz system as a special case but also a number of other common chaotic systems,

Case N:
$$\begin{array}{lll}
\dot{x} = a_1 x + a_2 y + a_3 z & u = \dot{x} & \\
\dot{y} = a_4 x + a_5 y + a_6 z + a_7 xz & v = \dot{y} & (4.7) \\
\dot{z} = a_8 x + a_9 y + a_{10} z + a_{11} xy + a_{12} y^2 & w = \dot{z}. &
\end{array}$$

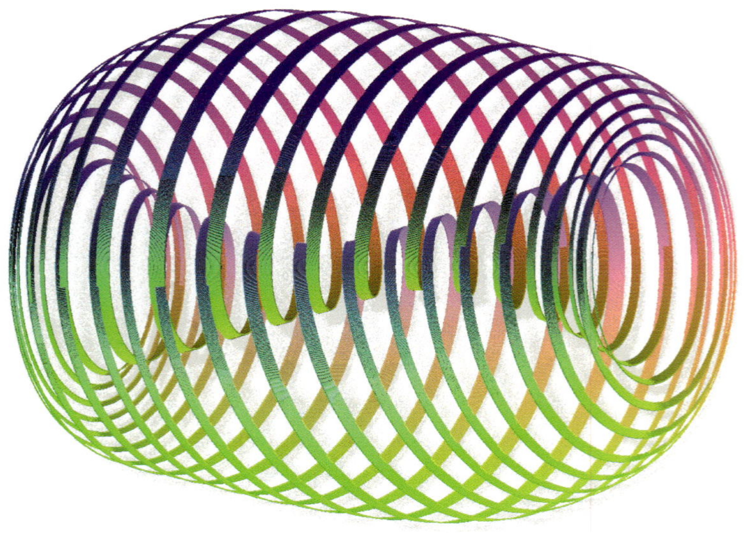

Fig. 4.11   `MLKJSFXIATKJG` $(PF = 0.417, CP = 0.891, LE = 0, DF = 2)$.

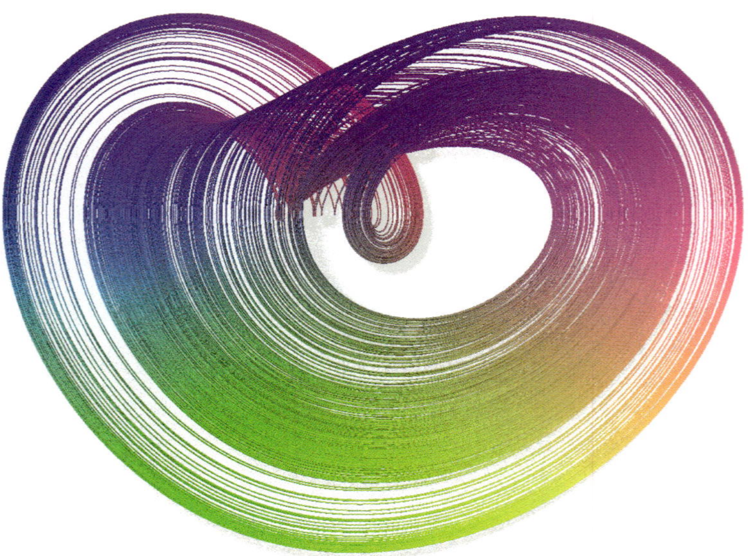

Fig. 4.12   `MKSYFQWKYJDSA` $(PF = 0.515, CP = 0.596, LE = 0.0329, DF = 2.1486)$.

Of course restricting the parameters to 26 values in the range $-1.2 \leq a \leq 1.3$ does not guarantee that you can obtain an attractor resembling the familiar the Lorenz butterfly, but in fact, it does suffice.

However, before examining such an example, look at Figs. 4.13 and 4.14 that show two elegant solutions of Eq. (4.7) with codes prefaced by N.

4.6 The Lorenz System

Buried in the previous general Eq. (4.7) are systems with the same form as those that are well known and that have been extensively studied provided you rescale them so that they can be represented using our restrictive coding scheme and retain the correct signs. For example, the Lorenz system in Eq. (4.6) can be written with different coefficients as

$$\dot{x} = y - 0.5x$$
$$\dot{y} = -xz + x - 0.1y \qquad (4.8)$$
$$\dot{z} = xy - 0.1z.$$

The coefficients have been chosen to give a Kaplan–Yorke dimension that is close to the value of 2.06215 [Sprott (2003)] for the classic case in Eq. (4.6) and thus with a similar attractor.

Equation (4.8) is represented by the code NHWMWLMCMMLWM with an attractor as shown in Fig. 4.15, which should look familiar because it has become an emblem of chaos and the 'butterfly effect.' The origin of that term is from the title of a talk given by Lorenz in 1972 at a meeting of the American Association for the Advancement of Science, 'Predictability: Does the Flap of a Butterfly's Wing in Brazil Set off a Tornado in Texas?' [Lorenz (1993)].

The Lyapunov exponent of $LE = 0.0462$ is not expected to agree with the value of 0.9056 using the conventional (σ, r, b) parameters because time has been rescaled to allow for the simpler coefficients, but the Kaplan–Yorke dimension should be and is approximately correct. For any dynamical system, the time can be rescaled without changing the appearance of the attractor since it only affects the rate at which the orbit moves about on the attractor.

Closely related to the Lorenz system is the Chen system [Chen and Ueta (1999)] which differs from it only in the signs of the x and y terms in the $\dot{y}$ equation,

$$\dot{x} = a(y - x)$$
$$\dot{y} = (c - a)x - xz + cy \qquad (4.9)$$
$$\dot{z} = xy - bz,$$

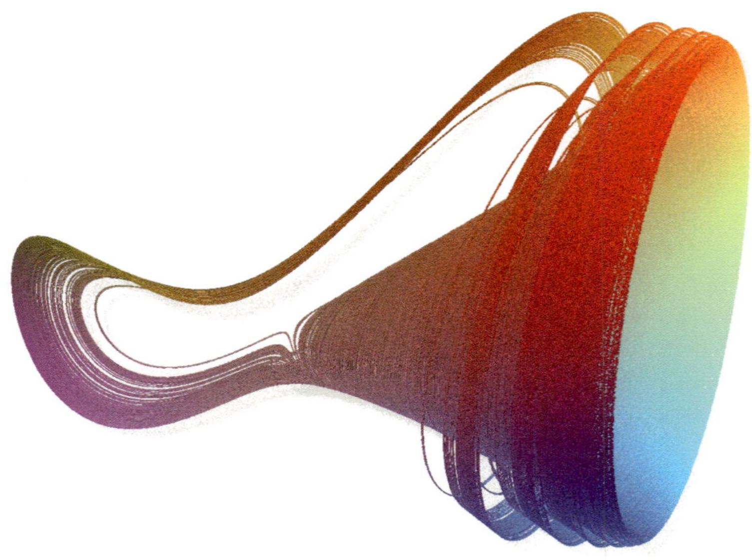

Fig. 4.13   NKURMTYTKPDAP ($PF = 0.395, CP = 0.897, LE = 0.0667, DF = 2.1429$).

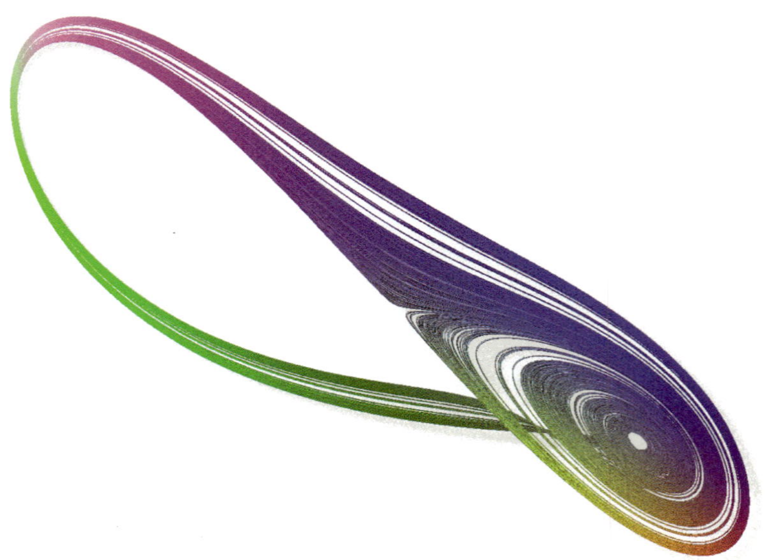

Fig. 4.14   NEHMKTLHQJHTA ($PF = 0.183, CP = 0.813, LE = 0.0222, DF = 2.0357$).

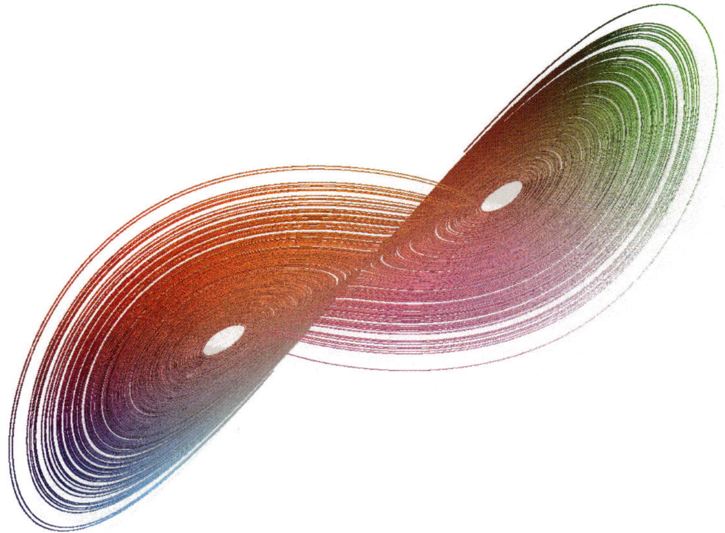

Fig. 4.15   `NHWMWLMCMMLWM` ($PF = 0.278, CP = 0.626, LE = 0.0463, DF = 2.0621$).

and which has chaotic solutions for $(a, b, c) = (35, 3, 28)$. In fact, it can be viewed as a special case of the Lorenz system with $r + \sigma = -1$ and time reversed [Algaba *et al.* (2013); Sprott (2015)]. Its attractor is not sufficiently different or elegant to warrant a figure.

4.7   The Rössler Prototype-4 System

When Lorenz (1963) published Eq. (4.6), it was the simplest known example of a three-dimensional autonomous system of ordinary differential equations that had a chaotic attractor since it has only seven terms and two quadratic nonlinearities, and that remained the case for thirteen years until Otto Rössler (1976) published a case with seven terms and a single quadratic nonlinearity. Unfortunately, his system is not in the form of any of the systems described so far because it contains a constant term, but three years later, he published an even simpler example with six terms and a single quadratic nonlinearity and with a similar attractor called the *prototype-4 system* [Rössler (1979)],

$$\dot{x} = -y - z$$
$$\dot{y} = x \tag{4.10}$$
$$\dot{z} = a(y - y^2) - bz.$$

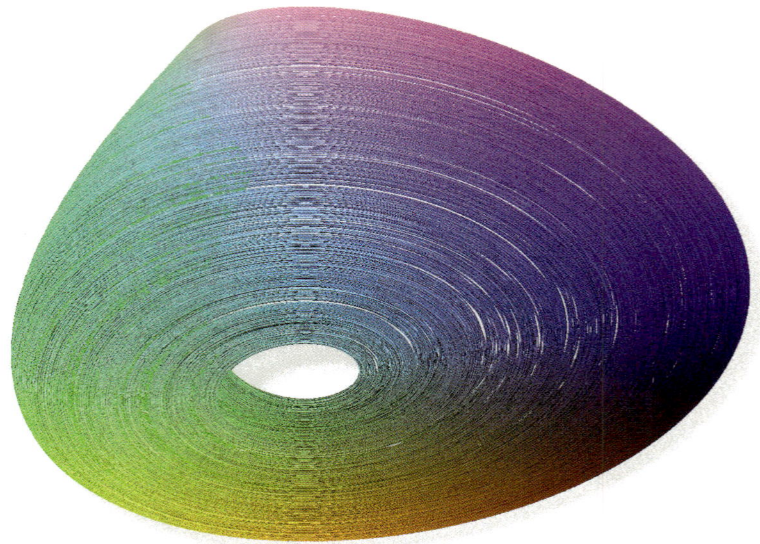

Fig. 4.16   `NMCCWMMMMRHMH` ($PF = 0.687, CP = 0.872, LE = 0.0938, DF = 2.1580$).

He showed values of the parameters (a, b) that give quasiperiodic $(0.2, 0)$ and chaotic $(0.386, 0.2)$ solutions, although the parameters $(0.5, 0.5)$ also give chaos [Sprott (2010)]. Conveniently, the latter parameters lend themselves to a representation in the form of Eq. (4.7) using the code `NMCCWMMMMRHMH` without the need for rescaling. The corresponding attractor shown in Fig. 4.16 is simple but not especially elegant.

Fifteen years later Sprott (1994a) published an additional fourteen examples of chaotic systems with six terms and one quadratic nonlinearity (plus four cases with five terms and two quadratic nonlinearities). Four of those cases (`G`, `I`, `J`, and `Q`) are in the form of Eq. (4.7), but their attractors resemble the one in Fig. 4.16 and thus will not be shown.

4.8   The Diffusionless Lorenz System

Many other systems can be put into the form of Eq. (4.7) by a transformation of variables. For example, consider the *diffusionless Lorenz system* [van der Schrier and Maas (2000); Munmuangsaen and Srisuchinwong

(2009)] which contains a constant term R,

$$\dot{x} = y - x$$
$$\dot{y} = -xz \qquad (4.11)$$
$$\dot{z} = xy - R,$$

and that has chaotic solutions for $R = 1$ but with a maximum Kaplan–Yorke dimension of 2.2354 at $R = 3.4693$ [Sprott (2007)]. Under the transformation $x \to x + \sqrt{R}$ and $y \to y + \sqrt{R}$, Eq. (4.11) becomes

$$\dot{x} = y - x$$
$$\dot{y} = -\sqrt{R}z - xz \qquad (4.12)$$
$$\dot{z} = \sqrt{R}x + \sqrt{R}y + xy,$$

which is in the form of Eq. (4.7). For $R = 1$ it can be represented by the code NCWMMMCCWWMWM with an attractor as shown in Fig. 4.17.

4.9 The Malasoma System

We mention one other system that is in the form of the cubic jerk function in Eq. (4.3) because it is claimed to be the simplest chaotic system that is parity invariant [Malasoma (2000)],

$$\dot{x} = y$$
$$\dot{y} = z \qquad (4.13)$$
$$\dot{z} = -1.2z + 1.2xy^2 - 0.2x,$$

or in more compact jerk form, $\dddot{x} = -1.2\ddot{x} + 1.2x\dot{x}^2 - 0.2x$ where the parameters have been scaled to allow it to be represented as KKMYMMMM-MAMMM with an attractor as shown in Fig. 4.18.

It will be left as an exercise for you to find additional examples of previously published systems that can be put in the form of one of the previous systems of equations since that constitutes a digression from our quest for elegant fractals.

4.10 Elegant Equations

In the companion book, *Elegant Chaos: Algebraically Simple Chaotic Flows* [Sprott (2010)], hundreds of continuous-time chaotic systems are listed whose equations are elegant. The equation elegance was defined by calculating an inelegance IE whose value is minimized. The inelegance is defined as the sum of the number of parameters plus the number of digits in their values including the decimal point but ignoring any leading or trailing zeroes.

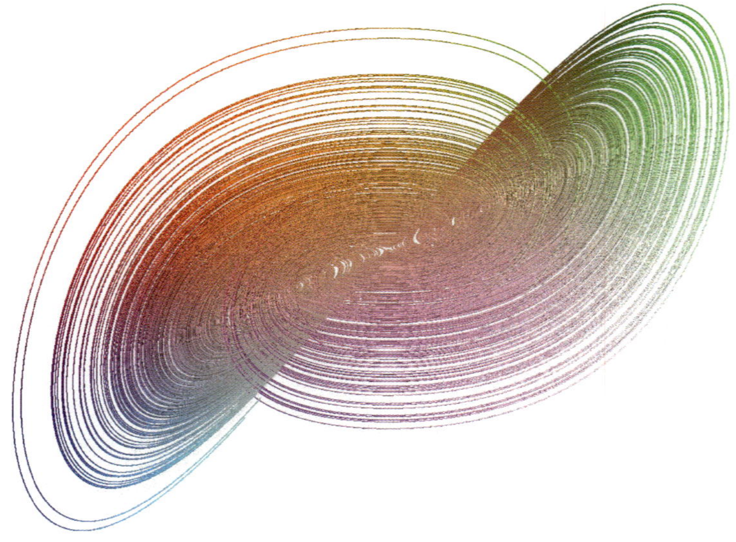

Fig. 4.17   NCWMMMCCWWMWM ($PF = 0.364, CP = 0.538, LE = 0.2101, DF = 2.1736$).

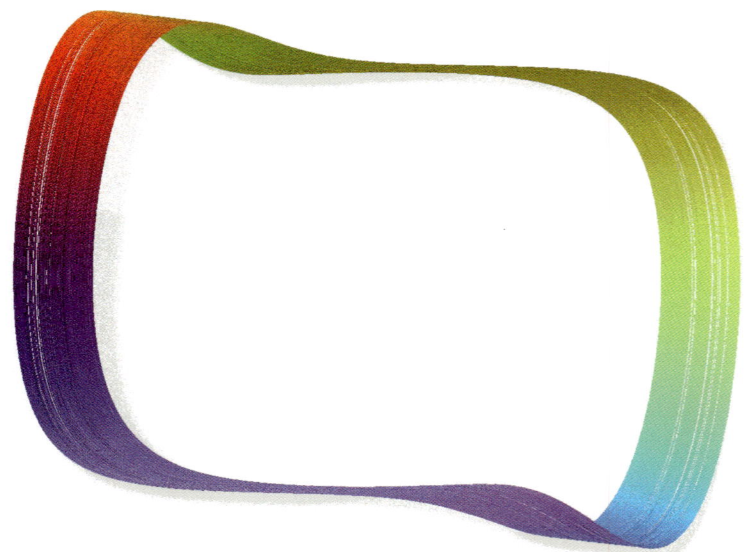

Fig. 4.18   KKMYMMMMMAMMM ($PF = 0.219, CP = 0.815, LE = 0.0298, DF = 2.0242$).

To see how this works, try calculating the value of IE for Eqs. (4.9) and (4.13). If you got $IE = 13$ for both cases, then you have the right idea. Note that constant, linear, and nonlinear terms are treated the same in calculating the inelegance and that the constraints on the parameters disallow integer values other than zero and ± 1.

The inelegance can also be calculated by looking at the code for each case, ignoring the prefix and suffix if there is one. There should be twelve such letters in all the preceding cases. Count the number of letters that are not M, then add 3 to that number for each A, B, X, Y, or Z, or add 2 for any other letter except C and W. That sum is the value of IE. If you want to practice, convince yourself that Fig. 4.15 has $IE = 13$ and Fig. 4.16 has $IE = 12$.

Thus it is relatively simple to minimize the equation inelegance for any of these systems by only considering parameter combinations for which IE is less than or equal to the best value (smallest IE) found so far. In this way, the following cases were determined to be the most elegant equations with chaotic solutions for the systems in this chapter:

Case J: JWWWMCMMMMWCM, $IE = 6$, Fig. 4.19

$$\begin{aligned}
\dot{x} &= (1 + x + y)y \\
\dot{y} &= -z \\
\dot{z} &= (x - y)x.
\end{aligned} \qquad (4.14)$$

Case K: KMMMCMCMMMMMC, $IE = 3$, Fig. 4.20

$$\begin{aligned}
\dot{x} &= y \\
\dot{y} &= z \\
\dot{z} &= -xz^2 - yx^2 - z^3.
\end{aligned} \qquad (4.15)$$

Case L: LKMMMMMCMMMWG, $IE = 8$, Fig 4.21

$$\begin{aligned}
\dot{x} &= y \\
\dot{y} &= z \\
\dot{z} &= u \\
\dot{u} &= -0.2x - z - \sin(0.6u).
\end{aligned} \qquad (4.16)$$

Case M: MKMMMMMCMMLWW, $IE = 9$, Fig. 4.22

$$\begin{aligned}
\dot{x} &= y \\
\dot{y} &= z \\
\dot{z} &= u \\
\dot{u} &= -0.2x - z - 0.1u + \tanh u.
\end{aligned} \qquad (4.17)$$

Case N: NMSMMMOMCMCMC, $IE = 9$, Fig. 4.23

$$\dot{x} = 0.6y$$
$$\dot{y} = 0.2z \tag{4.18}$$
$$\dot{z} = -x - z - y^2.$$

Equation (4.18) deserves special mention since it is in the form of the algebraically simplest dissipative system of ordinary differential equations that can exhibit chaos [Sprott (1997a)]. Making the transformation $(x, y, z, t) \rightarrow (0.36A^2x, 0.6Ay, 0.36A^3z, At)$ with $A = 1/\sqrt[3]{0.12} \approx 2.0274$ reduces Eq. (4.18) to the more familiar jerk form of $\dddot{x} = -A\ddot{x} - \dot{x}^2 - x$.

Look at Figs. 4.19 to 4.23 and decide whether you think elegant equations produce elegant strange attractors, or as what seems more likely, that the elegance of the equations has little bearing on the elegance of their attractors.

These elegant systems can also be displayed as icons, and Fig. 4.24 shows the same case as in Fig. 4.23 (the algebraically simplest chaotic flow) but displayed as a symmetric icon at higher resolution and in a larger size, except with the sign of x reversed to put the intricate structure away from the origin rather than near it. Perhaps this adds a bit of interest to what is otherwise a rather bland image.

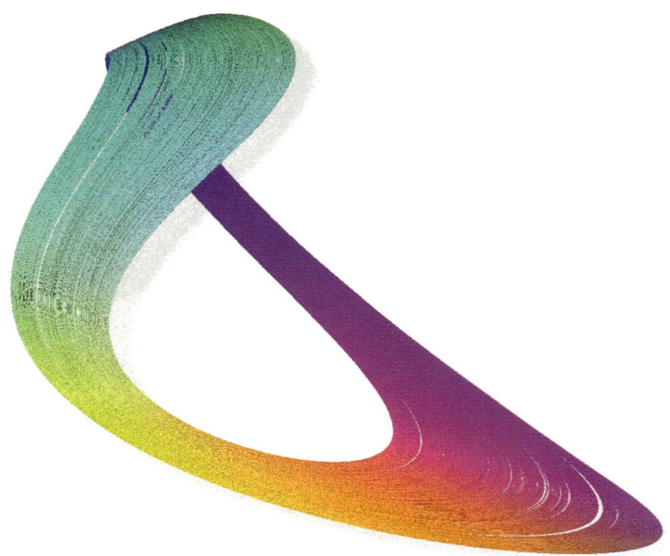

Fig. 4.19   JWWWMCMMMMWCM ($PF = 0.262, CP = 0.943, LE = 0.0334, DF = 2.0689$).

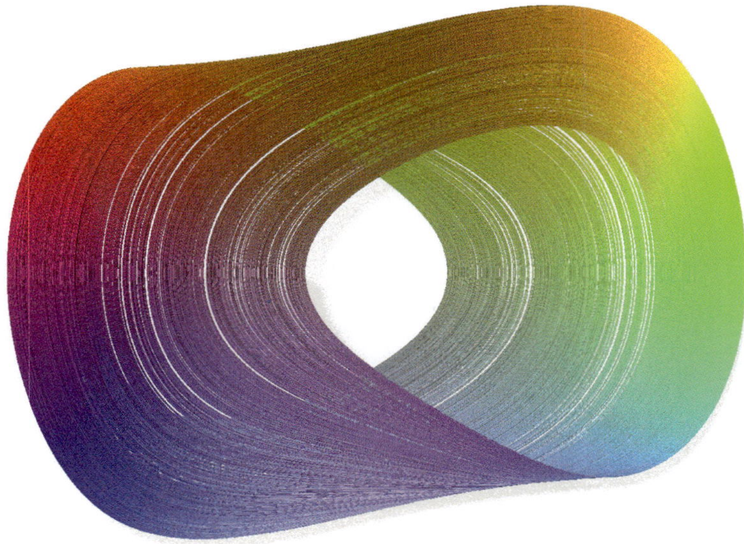

Fig. 4.20 KMMMCMCMMMMMC ($PF = 0.685, CP = 0.851, LE = 0.0911, DF = 2.0832$).

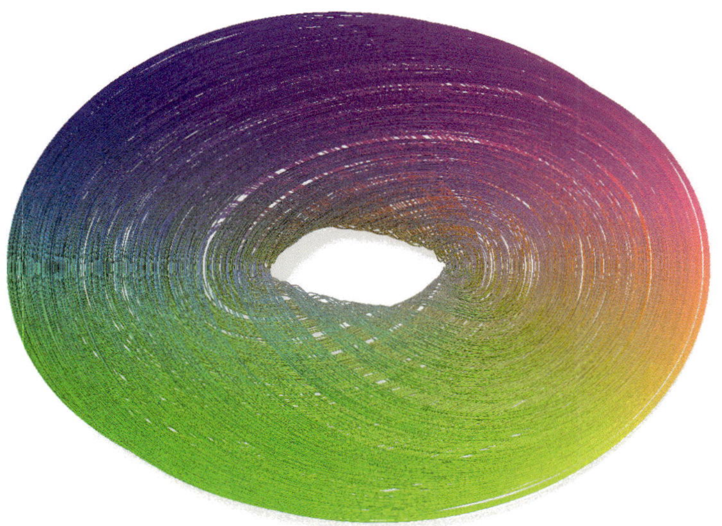

Fig. 4.21 LKMMMMMCMMMWU ($PF = 0.614, CP = 0.789, LE = 0, DF = 2$).

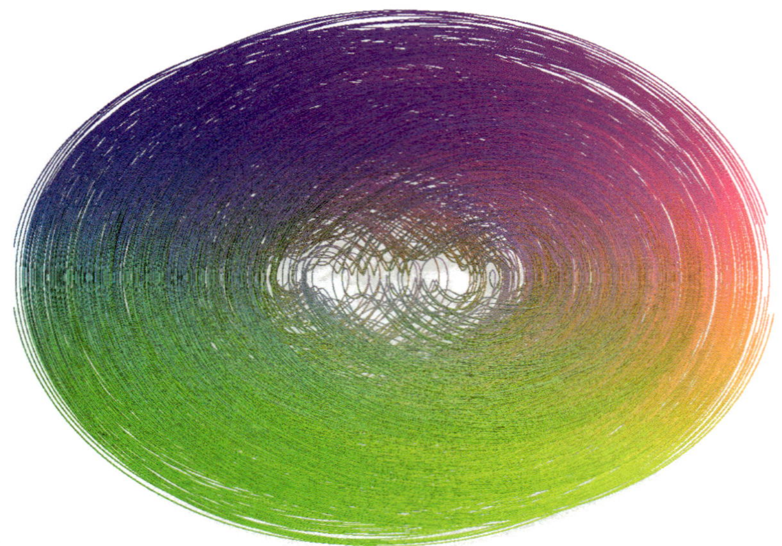

Fig. 4.22   MKMMMMMCMMLWW ($PF = 0.695, CP = 0.750, LE = 0.0065, DF = 3.3106$).

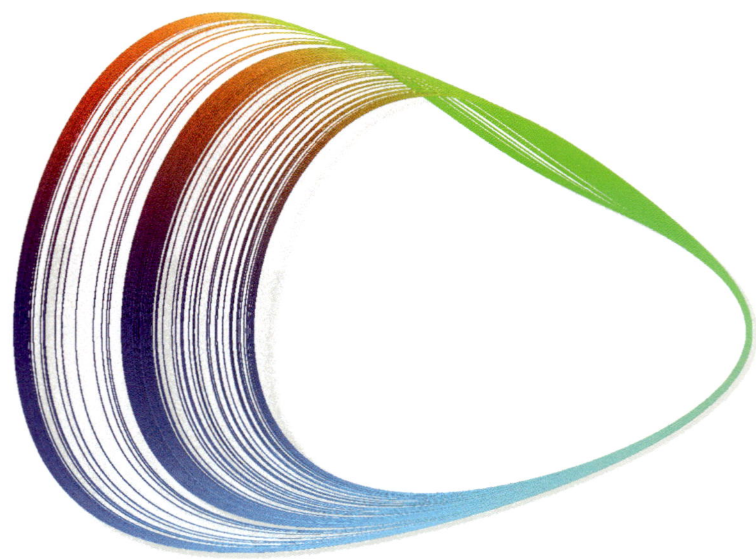

Fig. 4.23   NMSMMMOMCMCMC ($PF = 0.200, CP = 0.597, LE = 0.0177, DF = 2.0174$).

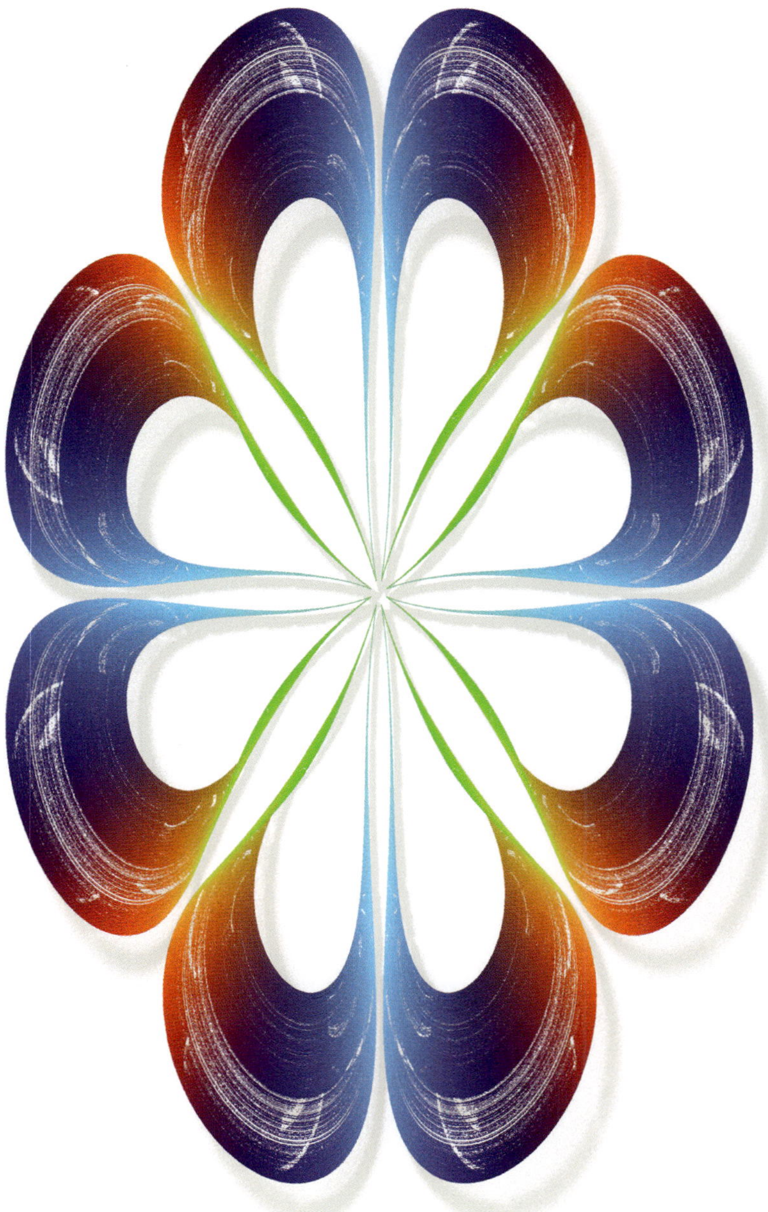

Fig. 4.24   The same case as in Fig. 4.23 but displayed as a symmetric icon.

Chapter 5

Iterated Function Systems

One of the most common and versatile methods for producing fractals makes use of iterated function systems. They are a kind of iterated map, and you can use the methods developed earlier to produce the images and to automatically select ones that are likely to be elegant from the vast majority that are completely dull. This chapter will show some such examples and will suggest ways to extend the method.

5.1 Affine Transformations

Iterated function systems were first studied by Hutchinson (1981) and were popularized by Barnsley (1988). The simplest type of iterated function system makes use of *affine transformations*, which in two-dimensions take the form

$$x_{n+1} = a_1 x + a_2 y + a_3$$
$$y_{n+1} = a_4 x + a_5 y + a_6. \tag{5.1}$$

Since these equations are linear, they have a single *fixed point* given by

$$x^* = \frac{a_2 a_6 - a_3(a_5 - 1)}{(a_1 - 1)(a_5 - 1) - a_2 a_4}$$
$$y^* = \frac{a_4 a_3 - a_6(a_1 - 1)}{(a_1 - 1)(a_5 - 1) - a_2 a_4} \tag{5.2}$$

that can be either stable or unstable (an attractor or a repellor). The unstable solutions are unbounded and approach infinity upon repeated iteration. The stable solutions approach the *point attractor* at the fixed point. Neither case produces an interesting fractal, and chaotic solutions are impossible because the equations are linear.

You can also think of the map in Eq. (5.1) as taking a set of points in the (x, y) plane and moving them all to new locations in the plane, generally with size scaling, translation, rotation, reflection, and shear. Stable solutions necessarily cause the area occupied by the points to contract with each

103

iteration. The amount of area contraction is determined by the magnitude of the determinant of the Jacobian matrix given by $|\det J| = |a_1 a_5 - a_2 a_4|$ which is the ratio of the area after the contraction to the area before, and hence it should be less than 1.0. A negative value of $\det J$ means that the region upon alternate iterations is reflected as if in a mirror in addition to any other distortions that may occur. Think of the succession of ever smaller images of yourself that you may have seen while sitting in the chair of a barber shop with mirrors on the wall in front of and behind you.

Note that area contraction does not guarantee boundedness since a set of points can continually contract in one direction while expanding in another, approaching a thin filament of zero area and infinite length. On the other hand, if $|\det J|$ is greater than 1.0, you can be sure that the solution is unbounded since the area continually increases without limit, and hence it will not produce a visually interesting solution.

Now suppose that in addition to the map with a stable fixed point that there is a second affine map with different parameters and hence a different stable fixed point. After some number of iterations of the first map, you switch to the second map causing the iterates to leave the vicinity of the first fixed point and approach the second, whereupon you switch back to the first map after a few iterations. The two fixed points compete for the orbit, typically producing a fractal attractor after many random switches between the two maps.

This process is called the *random iteration algorithm*, although the resulting pattern is anything but random and could equally well be produced by systematically employing all possible sequences of the two maps with all possible initial conditions on the resulting attractor. Note that the two fixed points lie on or immediately adjacent to the attractor because the random sequence will eventually exhibit an arbitrarily large number of iterations between switches, causing the orbit to asymptotically approach one of the fixed points. However, the location of the fixed points is usually not evident in the fractal pattern.

The extension of the method to more numerous and more complicated maps in dimensions higher than two is straightforward and allows such a wide variety of patterns that it can be used as a compression technique for photographic images [Barnsley and Hurd (1993)].

5.2 Two-dimensional Linear Systems

As an example of an iterated function system, consider the four two-dimensional affine maps given by

Case 0:

$$\text{Map 1: } x_{n+1} = a_1 + a_2 x + a_3 y$$
$$\text{Map 2: } x_{n+1} = a_4 + a_5 x + a_6 y$$
$$\text{Map 3: } x_{n+1} = a_7 + a_8 x + a_9 y$$
$$\text{Map 4: } x_{n+1} = a_{10} + a_{11} x + a_{12} y$$
$$y_{n+1} = x$$
$$z_{n+1} = y$$
$$u_{n+1} = z$$
$$v_{n+1} = u$$
$$w_{n+1} = v. \tag{5.3}$$

The remaining four variables (z, u, v, w) are slave variables that do not participate in the dynamics but are previous iterates of the dynamical variables (x, y), the last three of which (u, v, w) are used to determine the color of each point. The system is analogous to the two-dimensional scalar time-delay maps in Chapter 3.

One of the four maps is chosen randomly at each iteration using a random number r in the range of 0 to 1 and generating from it an integer $\lceil 4r \rceil$ in the range of 1 to 4 where $\lceil ... \rceil$ is the *ceiling function* (the smallest integer greater than its argument). The resulting system is a restricted form of an iterated function system but is capable of producing an enormous variety of patterns, some examples of which are in Figs. 5.1 to 5.4 with codes prefaced by an 0 to indicate that they came from Eq. (5.3). Note that some of these images are displayed as symmetric icons as indicated by the trailing digit in their code.

The figure captions in this chapter do not include a value for the Lyapunov exponent or Kaplan–Yorke dimension, which are quantities that are not usually calculated for iterated function systems. There are two ways to calculate a Lyapunov exponent when using the random iteration algorithm. You can either use the same random numbers for the primary and satellite orbits or different random numbers. Using the same numbers will typically give a negative value since the individual maps are contracting [Sprott (1994b)], while using different random numbers will give a large and meaningless positive value that is mostly governed by the randomness of the algorithm and hence has little to do with the character of the image.

Similarly, the Kaplan–Yorke dimension has little meaning since it is determined from the spectrum of Lyapunov exponents. However, the correlation dimension $D2$ is a measure of the space-filling character of the image, and so approximate values are given for it in the figure captions.

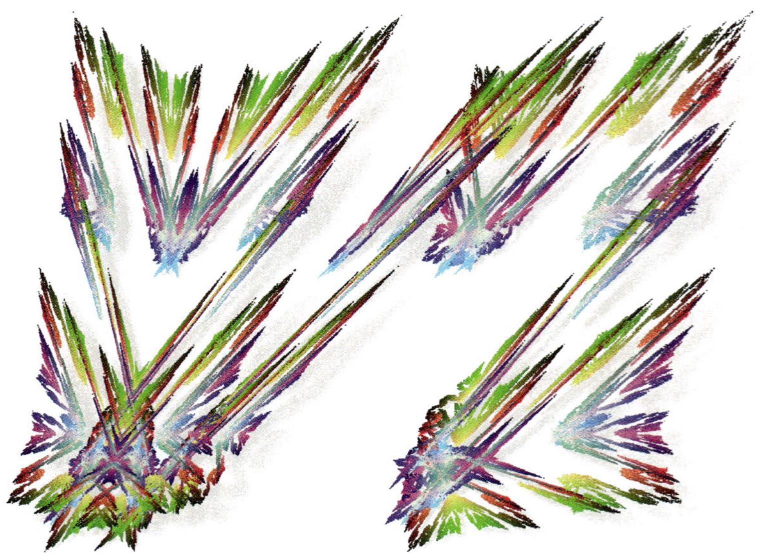

Fig. 5.1   OFNKMSNXNQHPO $(PF = 0.288, CP = 0.609, D2 = 1.743)$.

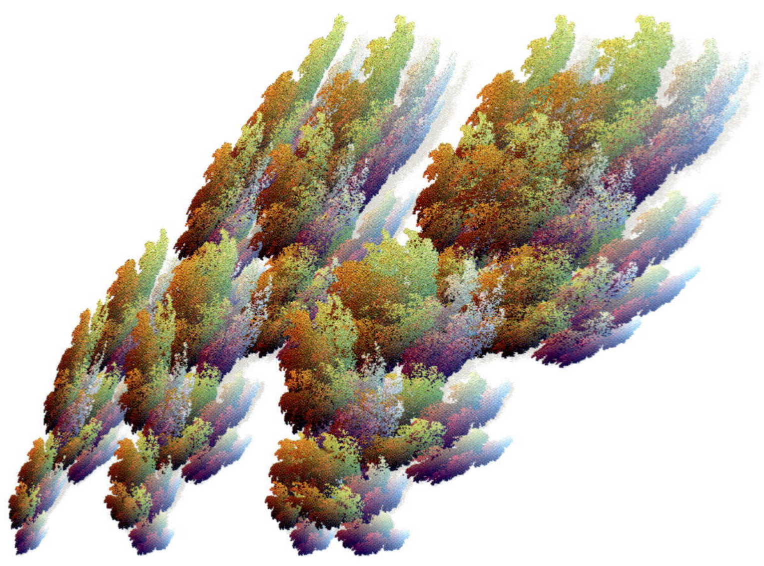

Fig. 5.2   OVSJMSJZUHGSK $(PF = 0.486, CP = 0.913, D2 = 1.962)$.

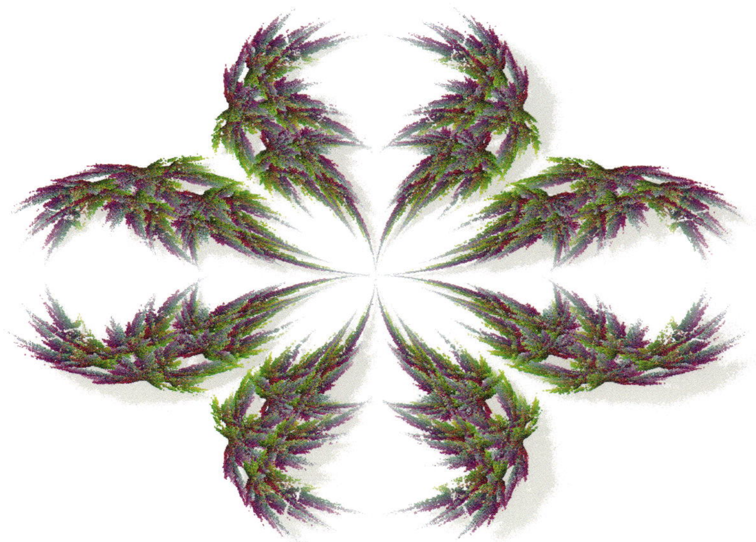

Fig. 5.3   `OBPRMNPBMNNHP8` $(PF = 0.262, CP = 0.729, D2 = 1.884)$.

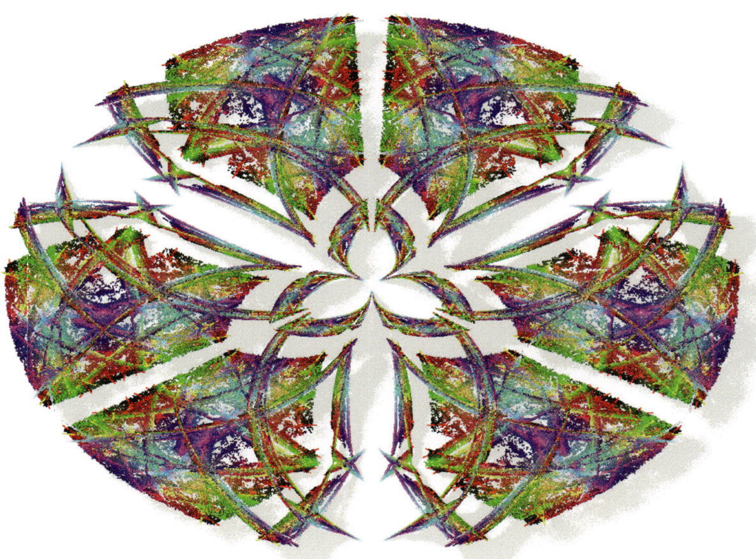

Fig. 5.4   `OZENWPGGONYMQ6` $(PF = 0.458, CP = 0.629, D2 = 1.745)$.

5.3 Three-dimensional Linear Systems

An example of a three-dimensional iterated function system with three affine maps is

$$
\begin{aligned}
&\text{Case P:}\\
&\text{Map 1: } x_{n+1} = a_1 + a_2 x + a_3 y + a_4 z\\
&\text{Map 2: } x_{n+1} = a_5 + a_6 x + a_7 y + a_8 z\\
&\text{Map 3: } x_{n+1} = a_9 + a_{10} x + a_{11} y + a_{12} z\\
&y_{n+1} = x\\
&z_{n+1} = y\\
&u_{n+1} = z\\
&v_{n+1} = u\\
&w_{n+1} = v.
\end{aligned}
\tag{5.4}
$$

The variables (u, v, w) are slave variables that do not participate in the dynamics but are used to determine the color of each point. Sample images produced by Eq. (5.4) are shown in Figs. 5.5 to 5.8 with codes prefaced by a P.

5.4 Two-dimensional Quadratic Systems

Iterated function systems can also employ nonlinear transformations, of which there are countless possibilities. A simple example of a two-dimensional iterated function system with two quadratic maps is

$$
\begin{aligned}
&\text{Case Q:}\\
&\text{Map 1: } x_{n+1} = a_1 + a_2 x + a_3 y + a_4 x^2 + a_5 xy + a_6 y^2\\
&\text{Map 2: } x_{n+1} = a_7 + a_8 x + a_9 y + a_{10} x^2 + a_{11} xy + a_{12} y^2\\
&y_{n+1} = x\\
&z_{n+1} = y\\
&u_{n+1} = z\\
&v_{n+1} = u\\
&w_{n+1} = v.
\end{aligned}
\tag{5.5}
$$

The variables (z, u, v, w) are slave variables that do not participate in the dynamics. Sample images produced by Eq. (5.5) are shown in Figs. 5.9 to 5.12 with codes prefaced by a Q. Since the maps are nonlinear, they can be individually chaotic, and the resulting images combine the features of strange attractors and iterated function systems and are thus enormously diverse and especially elegant.

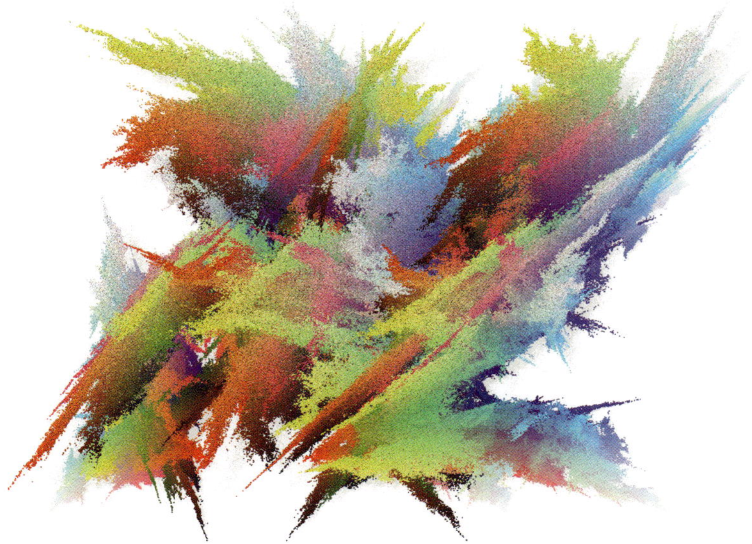

Fig. 5.5   PGTIPTKIJXQLJ ($PF = 0.499, CP = 0.892, D2 = 2.655$).

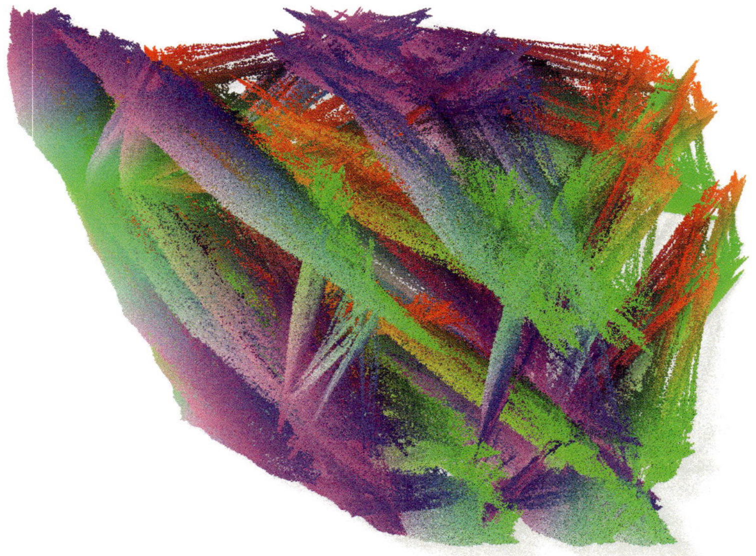

Fig. 5.6   PHIQOJJKQJKSL ($PF = 0.565, CP = 0.916, D2 = 2.308$).

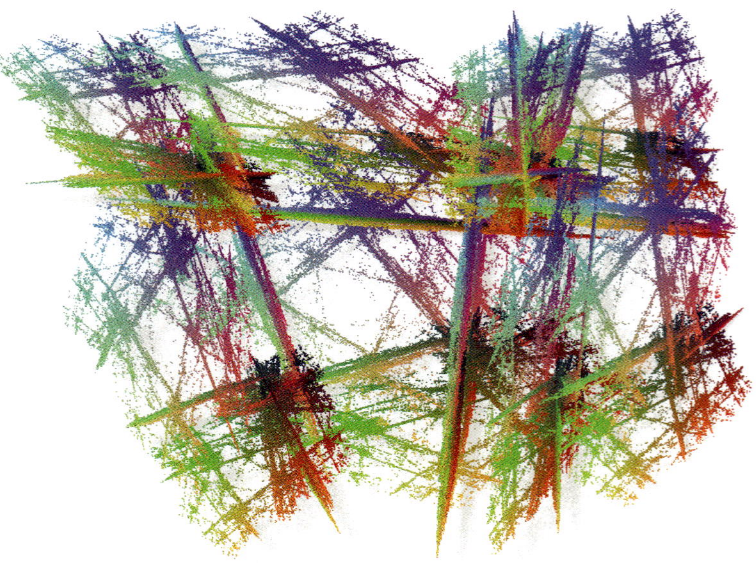

Fig. 5.7　PROMNYLILJKFN ($PF = 0.422, CP = 0.525, D2 = 1.800$).

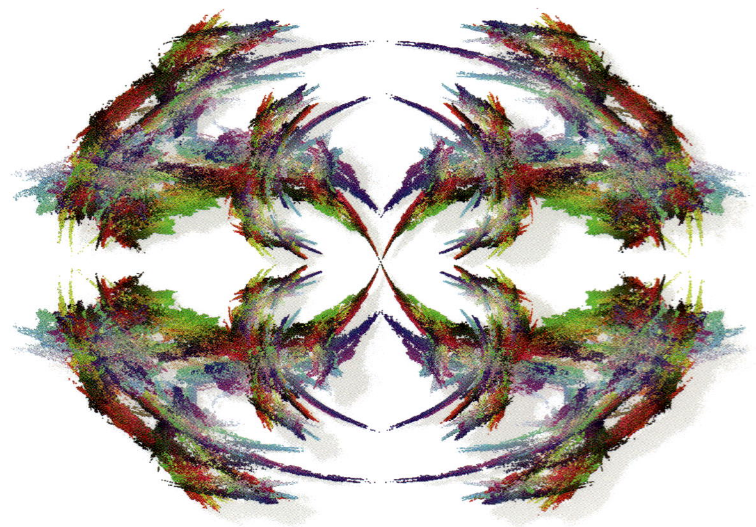

Fig. 5.8　PZKOOJUNHDHON4 ($PF = 0.357, CP = 0.763, D2 = 2.097$).

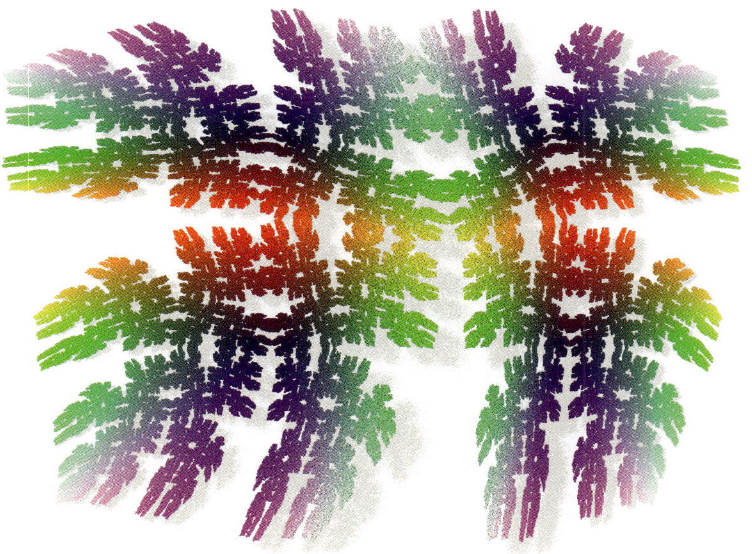

Fig. 5.9 **QLJHEDLLLTDVN** $(PF = 0.448, CP = 0.776, D2 = 1.819)$.

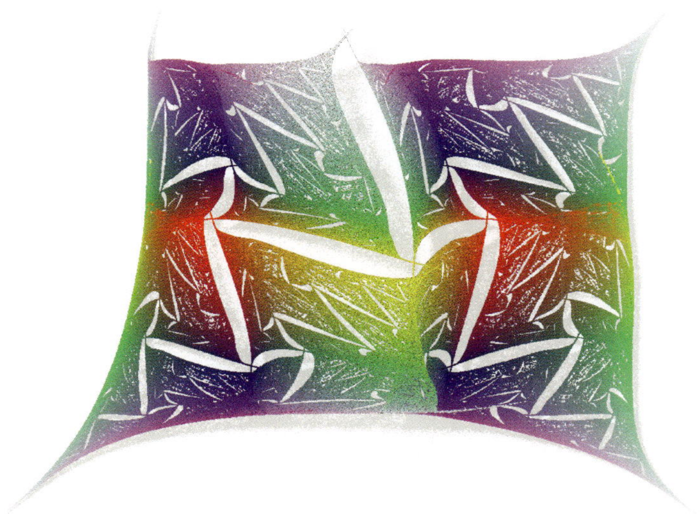

Fig. 5.10 **QMRPJFSURAFNU** $(PF = 0.330, CP = 0.602, D2 = 1.428)$.

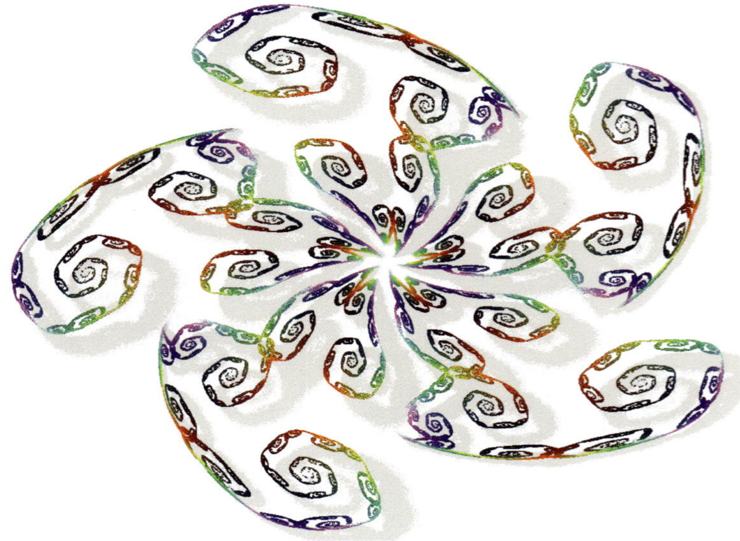

Fig. 5.11   `QLNPKNQPRLMMD5` $(PF = 0.127, CP = 0.269, D2 = 1.171)$.

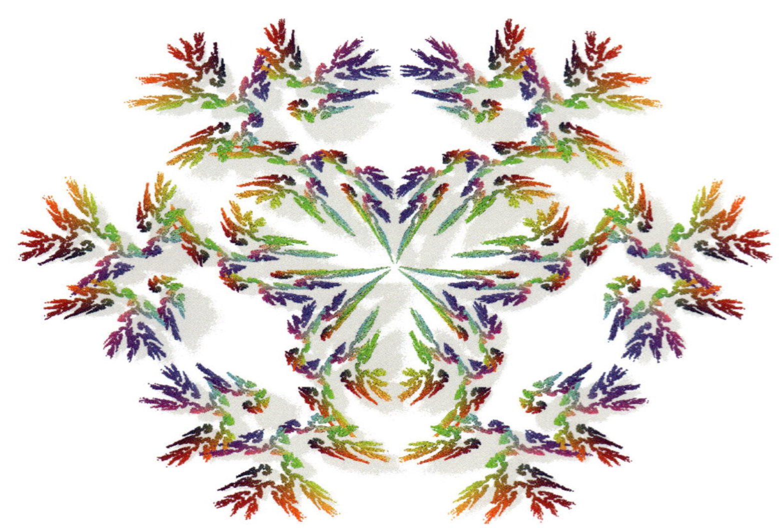

Fig. 5.12   `QRIEQRSPHKOOF6` $(PF = 0.195, CP = 0.472, D2 = 1.552)$.

5.5   Three-dimensional Cubic Systems

An example of a three-dimensional iterated function system with three cubic maps is

Case R:
Map 1: $x_{n+1} = a_1 + a_2 x^3 + a_3 y^3 + a_4 z^3$
Map 2: $x_{n+1} = a_5 + a_6 x^3 + a_7 y^3 + a_8 z^3$
Map 3: $x_{n+1} = a_9 + a_{10} x^3 + a_{11} y^3 + a_{12} z^3$

$$y_{n+1} = x \tag{5.6}$$
$$z_{n+1} = y$$
$$u_{n+1} = z$$
$$v_{n+1} = u$$
$$w_{n+1} = v.$$

The variables (u, v, w) are slave variables that do not participate in the dynamics. Sample images produced by Eq. (5.6) are shown in Figs. 5.13 to 5.16 with codes prefaced by an R.

5.6   Elegant Equations

As for the strange attractors in the previous chapter, we can ask what are the most elegant equations that lead to non-trivial iterated function systems. Some of the most elegant forms for each of the four previous systems are:

Case O: OMIHCMIMMOMTM, $IE = 16$, Fig. 5.17

$$\text{Map 1: } x_{n+1} = -0.4x - 0.5y$$
$$\text{Map 2: } x_{n+1} = -1 - 0.4y$$
$$\text{Map 3: } x_{n+1} = 0.2y \tag{5.7}$$
$$\text{Map 4: } x_{n+1} = 0.7x$$
$$y_{n+1} = x$$

Case P: PCPMMMMFQIMJM, $IE = 16$, Fig. 5.18

$$\text{Map 1: } x_{n+1} = -1 + 0.3x$$
$$\text{Map 2: } x_{n+1} = -0.7y + 0.4z$$
$$\text{Map 3: } x_{n+1} = -0.4 - 0.3y \tag{5.8}$$
$$y_{n+1} = x$$
$$z_{n+1} = y$$

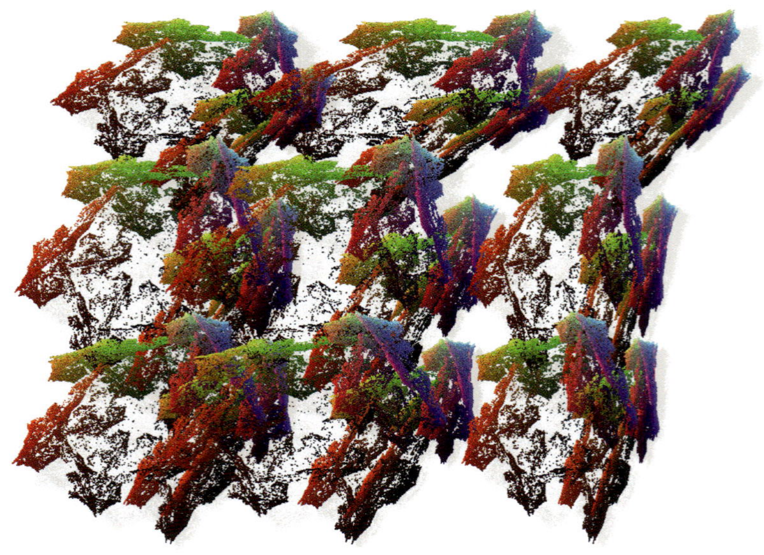

Fig. 5.13   RRUFJNQEGPVCG $(PF = 0.486, CP = 0.583, D2 = 1.249)$.

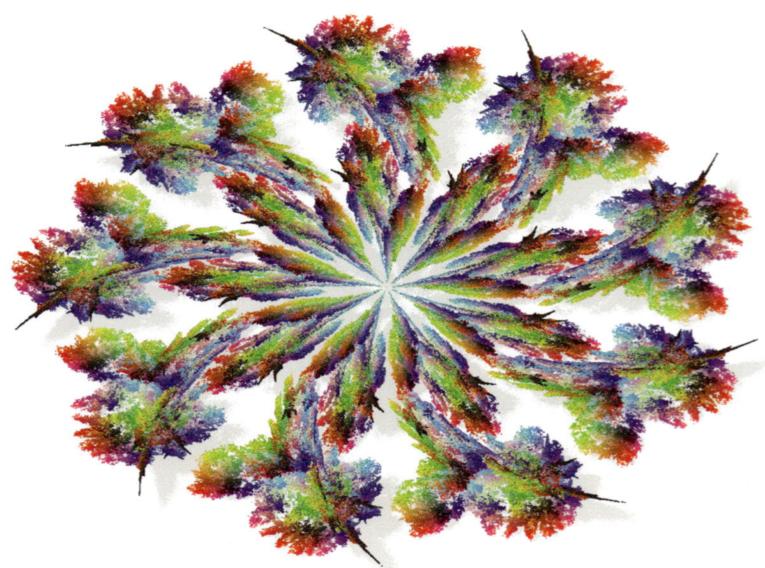

Fig. 5.14   RITFSIPGQJVPP9 $(PF = 0.462, CP = 0.649, D2 = 2.194)$.

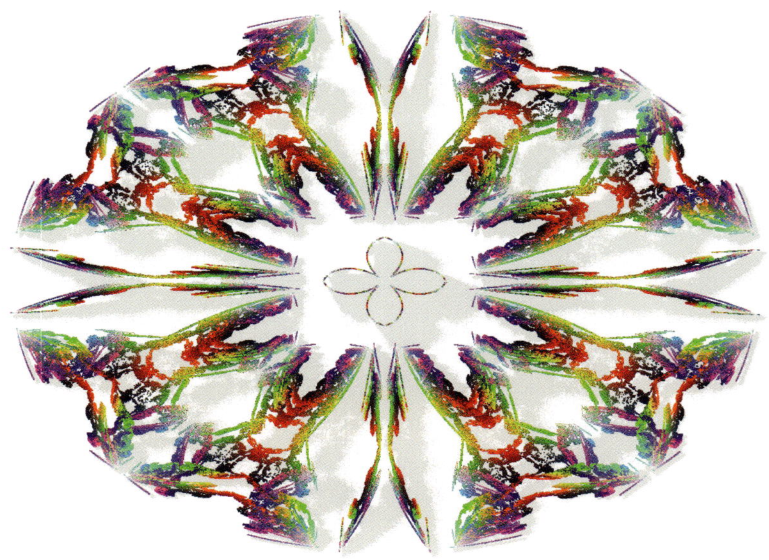

Fig. 5.15   ROQSXOGMMQMAG ($PF = 0.312, CP = 0.544, D2 = 0.821$).

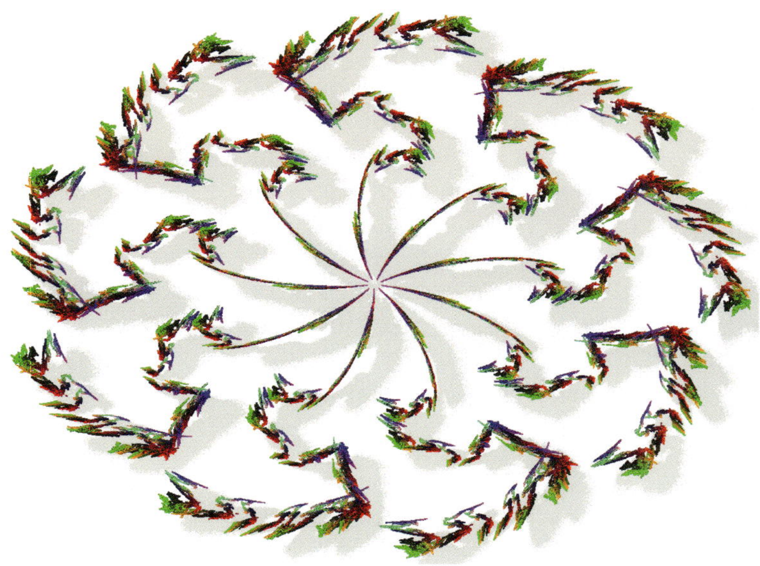

Fig. 5.16   RTKMORPKNQFLN9 ($PF = 0.158, CP = 0.560, D2 = 1.544$).

Case Q: QTMMMIMCCMRMQ, $IE = 14$, Fig 5.19

$$\text{Map 1: } x_{n+1} = 0.7 - 0.4xy$$
$$\text{Map 2: } x_{n+1} = -1 - x + 0.5x^2 + 0.4y^2 \tag{5.9}$$
$$y_{n+1} = x$$

Case R: RMMWCOMWWQROM, $IE = 16$, Fig 5.20

$$\text{Map 1: } x_{n+1} = y^3 - z^3$$
$$\text{Map 2: } x_{n+1} = 0.2 + x^3 - y^3 - z^3$$
$$\text{Map 3: } x_{n+1} = 0.4 + 0.5x^3 + 0.2y^3 \tag{5.10}$$
$$y_{n+1} = x$$
$$z_{n+1} = y$$

The attractors for these systems are shown in Figs. 5.17 to 5.20, respectively.

Here are some additional cases with elegant equations but whose solutions produce patterns that are somewhat less elegant and thus are not shown but that you might find appealing:

```
OMPLMMJWMOMFM  IE = 16  PF = 0.043  CP = 0.263  SK = 0.656
OMQMMTMMMIWRG  IE = 16  PF = 0.145  CP = 0.020  SK =-0.758
PMMVMGMMGWMDI  IE = 16  PF = 0.186  CP = 0.760  SK =-1.160
PMNMMPMMHWLMQ  IE = 16  PF = 0.510  CP = 0.430  SK =-0.201
PMTMICHMMRLMM  IE = 16  PF = 0.093  CP = 0.230  SK =-0.302
PNMQMWOMMMRMJ  IE = 16  PF = 0.078  CP = 0.399  SK = 0.926
PROMMWMMOMMKK  IE = 16  PF = 0.104  CP = 0.466  SK = 0.356
PSMMVMLSMCMMD  IE = 16  PF = 0.409  CP = 0.741  SK = 0.270
PTMOHMMMPCMMJ  IE = 16  PF = 0.701  CP = 0.874  SK =-0.679
QHMZMWWMMROWM  IE = 16  PF = 0.040  CP = 0.348  SK = 0.485
QLMCMHMMMCICD  IE = 15  PF = 0.721  CP = 0.932  SK = 0.181
QMLSMRMOWWMCC  IE = 16  PF = 0.504  CP = 0.838  SK = 1.172
QMWPLGMMMWMMA  IE = 15  PF = 0.042  CP = 0.682  SK =-1.164
QOMICCVMMQCMM  IE = 15  PF = 0.057  CP = 0.299  SK =-0.001
QOMSMMMGUMMWU  IE = 16  PF = 0.576  CP = 0.732  SK = 0.232
QWMUMCHWWMCEM  IE = 14  PF = 0.373  CP = 0.964  SK = 0.938
RJWCWMWMCHQMJ  IE = 17  PF = 0.370  CP = 0.757  SK =-0.636
```

5.7 Alternate Coloring Scheme

The method for coloring the previous images is only one of many possibilities. Another option is inspired by the *Fractals Colouring Book* [Sprott (2014)]. Suppose you plot the successive (x, y) iterates in black without

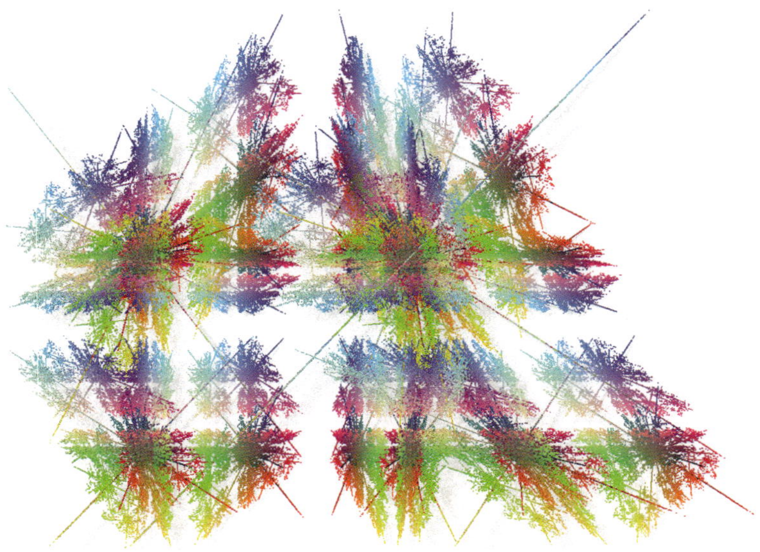

Fig. 5.17   OMIHCMIMMOMTM $(PF = 0.352, CP = 0.351, D2 = 0.791)$.

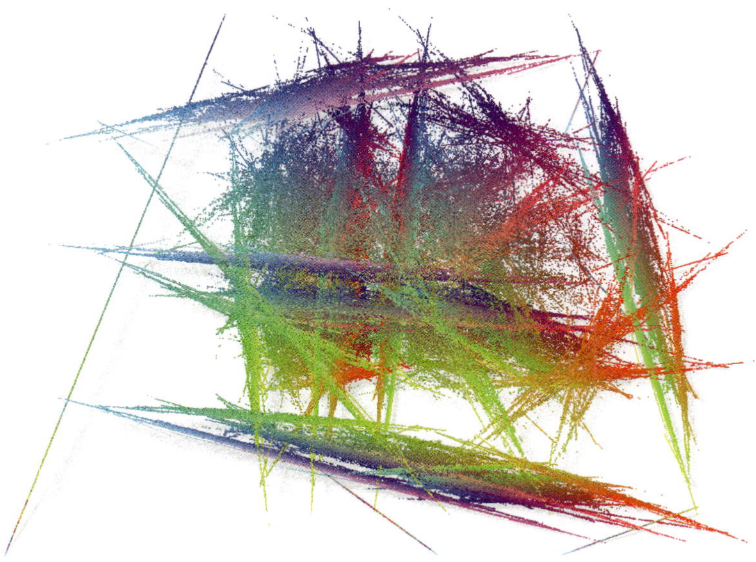

Fig. 5.18   PCPMMMMFQIMJM $(PF = 0.340, CP = 0.486, D2 = 0.968)$.

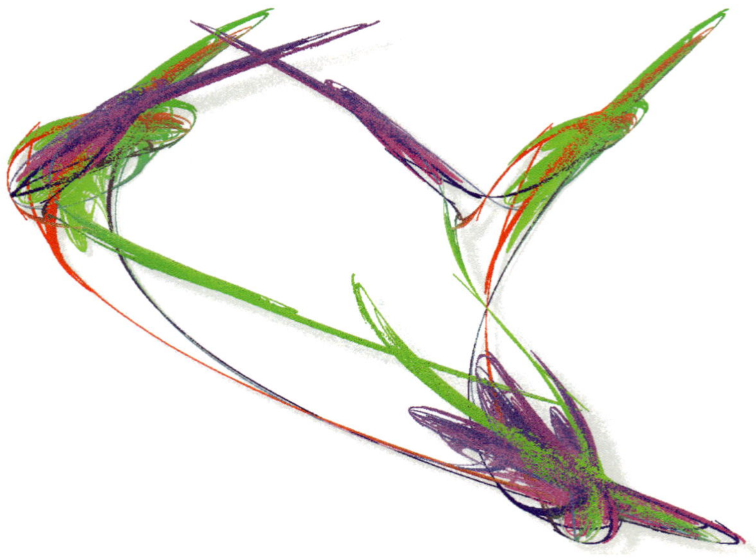

Fig. 5.19   QTMMMIMCCMRMQ ($PF = 0.150, CP = 0.763, D2 = 1.513$).

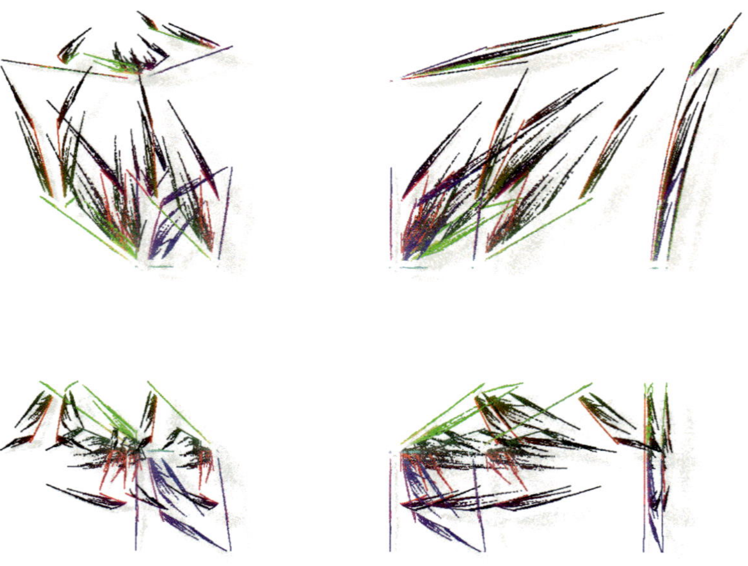

Fig. 5.20   RMMWCOMWWQROM ($PF = 0.102, CP = 0.277, D2 = 0.637$).

using any anti-aliasing. Then you color the white regions totally enclosed by black lines with one of a small number of randomly chosen colors as you would with a coloring book and a handful of crayons. The BASIC language has a PAINT statement that does this automatically. If you are using a different computer language that does not have a similar statement, you are on your own.

The number and choice of colors is arbitrary, but consider a palette of six colors including the usual *additive primaries* of red, green, and blue, and the *subtractive primaries* of cyan, magenta, and yellow. The former three are used when producing white light from three colors as is typical on computer monitors, and the latter three (usually along with black) are used when producing colors by selectively absorbing reflected white light using ink on a printed page. Thus these colors should look good on both a computer screen and in print, although somewhat different. Decreasing the number of colors from $256^3 = 16\,777\,217$ to 6 sounds like a reduction in elegance, but sometimes 'less is more,' especially when the chosen colors are vivid as in the current case.

Note that since the colors are chosen randomly, you will get a different rendition of the image every time unless you reseed the random number generator with the same seed each time the colors are chosen, and two programs will produce differently colored images unless they use the same algorithm for generating random numbers. More details about portable random number generators and choosing seeds will be given in Chapter 8.

The selection methods previously employed work well to eliminate the vast majority of uninteresting solutions. If you want to do better, you can keep track of the number of pixels with each of the seven colors (including black) and eliminating any cases in which more than about 90% of the pixels have the same color, or perhaps the same two colors since they probably have insufficient interesting structure.

Clearly this coloring method works best if the fractal is not too dense (a relatively small pixel fraction PF) and has moderately thin lines with some but not too many intersections. Figures 5.21 to 5.27 show some examples including some that are symmetric icons. Case Q in Eq. (5.5) seems especially good for producing such images. How would you rank their elegance relative to others in this chapter? Do you like them better with a colored or a white background? Why do you think Case Q makes especially good images with this coloring method, and would you expect the method to work well with images from the previous chapters?

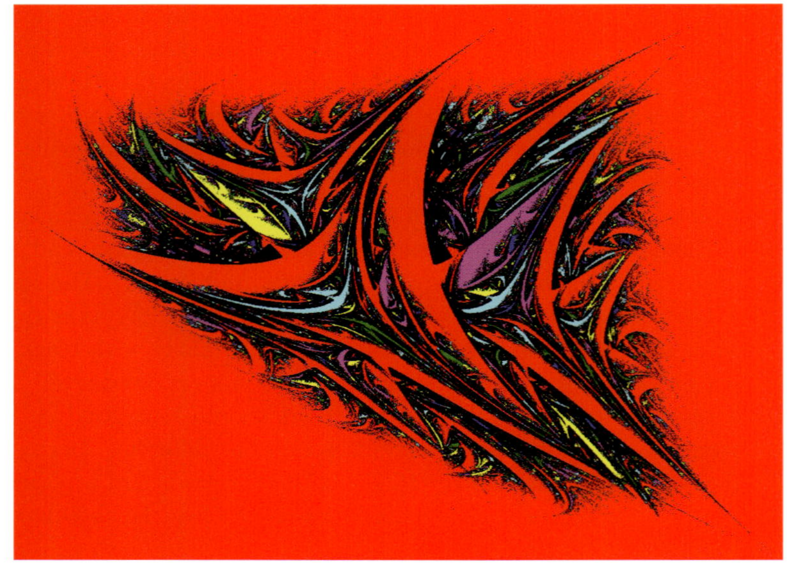

Fig. 5.21   QBVDYMJKVVUSU $(PF = 0.223, CP = 0.583, D2 = 1.391)$.

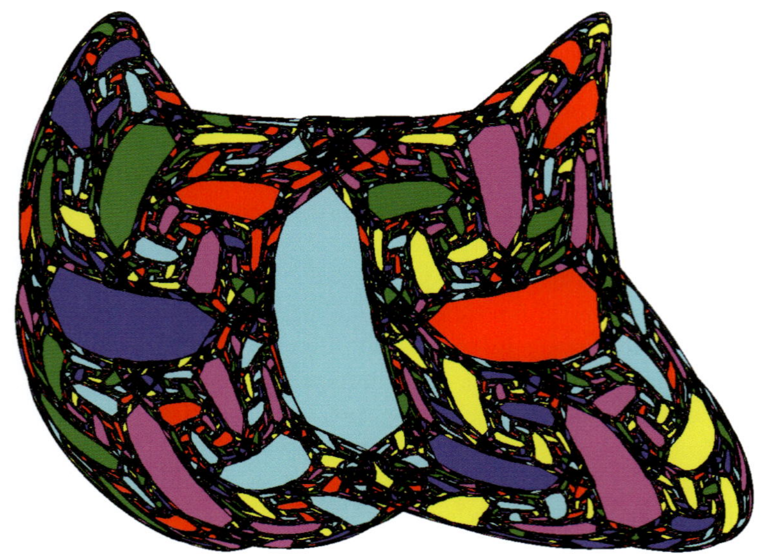

Fig. 5.22   QOIHXSJLKQYNS $(PF = 0.320, CP = 0.737, D2 = 1.717)$.

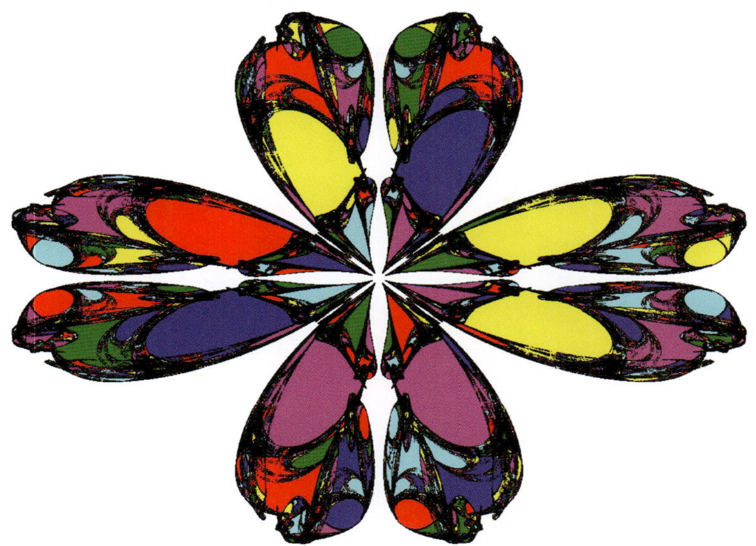

Fig. 5.23   `QGNQRPOREIAOD8` ($PF = 0.196, CP = 0.573, D2 = 1.191$).

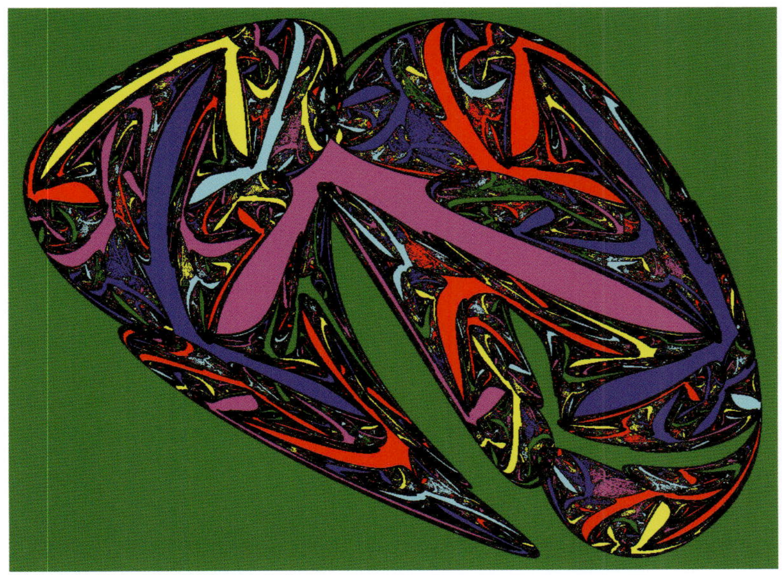

Fig. 5.24   `QPMSEEDKHIXKM` ($PF = 0.336, CP = 0.611, D2 = 1.304$).

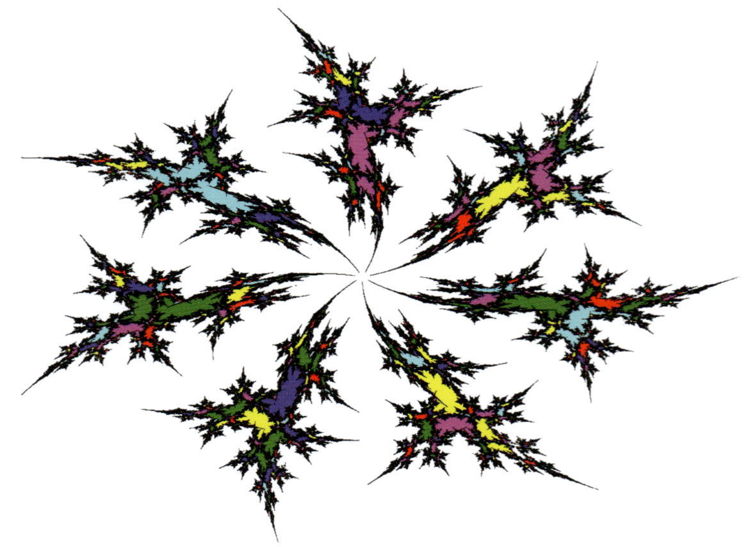

Fig. 5.25   QOPFNEZMPQGPL7 $(PF = 0.124, CP = 0.560, D2 = 1.675)$.

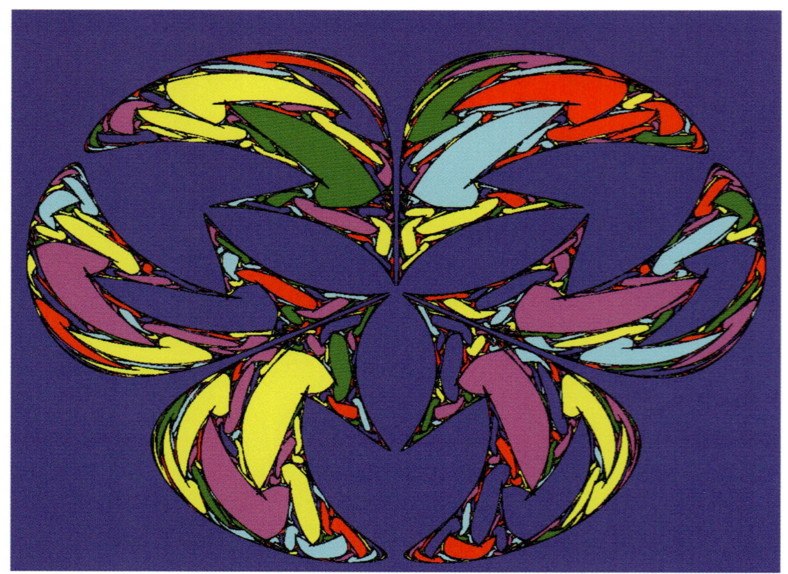

Fig. 5.26   QOMVHOEPJNOAK6 $(PF = 0.132, CP = 0.395, D2 = 1.420)$.

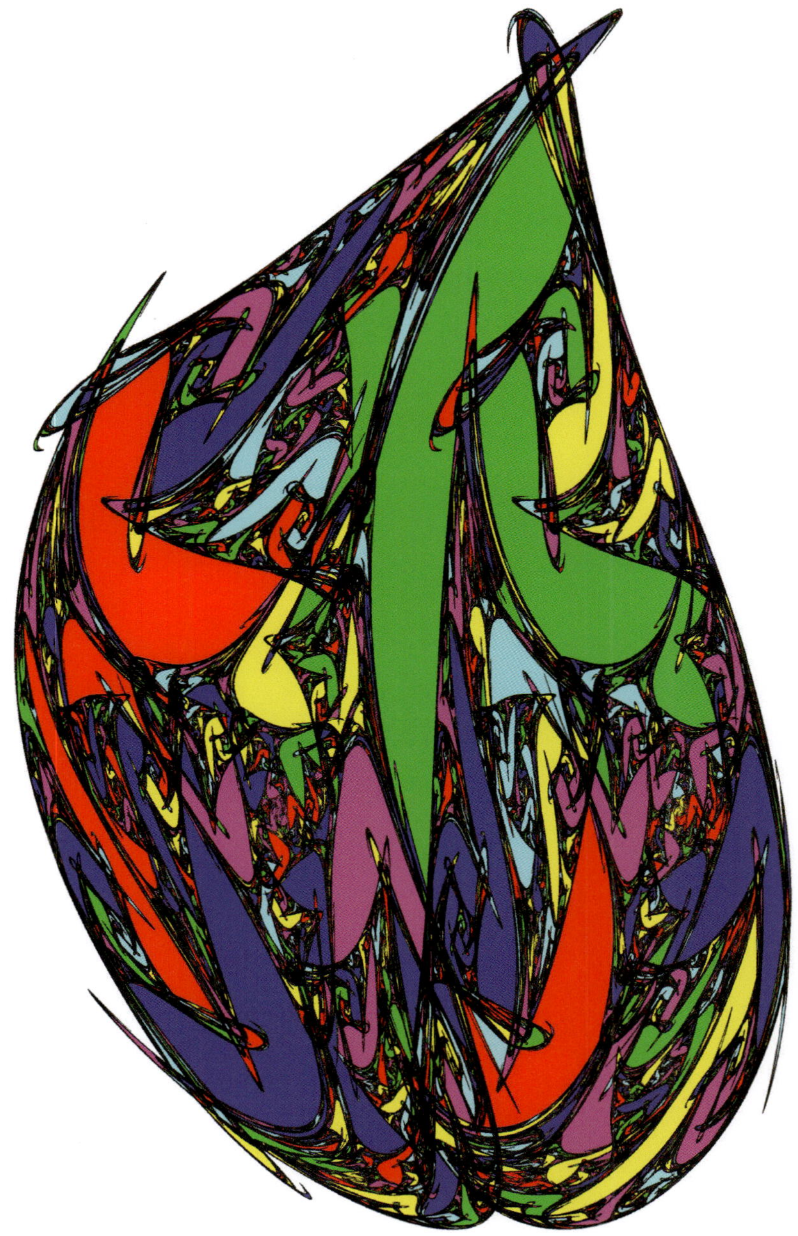

Fig. 5.27   QNNIKTKFUHWQL ($PF = 0.241, CP = 0.662, D2 = 1.556$).

Chapter 6

Escape-time Plots

Some of the most elegant fractals are produced by plots showing in color the number of iterations required for an orbit starting at a particular point to escape beyond some boundary surrounding an attractor. In fact, the famous Mandelbrot set and its associated Julia sets are the best known such examples. This chapter shows a number of other less familiar examples employing the maps and flows in the previous chapters as well as a special map that is a generalization of the usual Julia sets.

6.1 Basins of Attraction

Most attractors reside within a region of space that represents their basin of attraction within which all initial conditions are drawn to the attractor, and outside of which the orbits typically escape to infinity, although it is possible for there to be multiple attractors, each with its own non-overlapping basin of attraction. We already saw one such example for the Hénon map in Fig. 1.12. The boundary of the basin can be relatively smooth and simple as with the Hénon map, or it can be a complicated and elegant fractal.

The dimension of the basin is equal to the dimension of the space in which the attractor is embedded, which is usually the dimension of the dynamical system (the number of active independent variables in the governing equations). The dimension of the boundary of the basin is usually one less than the dimension of the basin, although it can be a non-integer in those frequent cases where it is a fractal. One-dimensional basins do not make elegant patterns, and basins with dimensions higher than two are difficult to plot except as a cross section in a particular plane with fixed values of the remaining variables. Thus we will consider only two-dimensional basins in the variables (x, y) with $z = u = \cdots = 0$.

125

Although the basin boundary can be an elegant fractal with a noninteger dimension less than or equal to the dimension of the basin, it is customary to color the region outside the basin according to the number of iterations required for the orbit to escape the vicinity of the attractor. The criterion used here will be $x^2 + y^2 > 100^2$. Thus if the orbit ever reaches a distance of 100 from the origin near which the attractor is located, it is assumed that it will never return to the attractor. There is no guarantee of this in general, although in certain cases a much smaller bound can be proved to be sufficient [Branner (1988)].

The procedure then is to choose initial conditions on a grid in $(x, y, 0, 0, \ldots)$ and plot each point of the grid in a color according to the number of iterations required for an orbit starting at that point to escape. For those orbits that do not escape, we take a 'bailout condition' of 1000 iterations after which the point is considered to be in the basin if it has not escaped, in which case that initial point is colored white. Most of the computation time is consumed by those orbits that do not escape or that lie near the basin boundary and require many iterations to escape. Other more advanced methods can be used to speed the calculation such as testing for periodicity of the orbit, but they are not necessary for the purposes of making elegant fractal patterns.

6.2   Color Selection

There are many ways to choose the colors for the escape-time contours from the $256^3 = 16\,777\,216$ available colors. One way is to specify the color according to the hue, saturation, and intensity (h, s, i), from which the red, green, and blue (r, g, b) intensities can be calculated according to the following:

$$
\begin{aligned}
&\text{For } h = 0 \text{ to } 2\pi/3 : \\
&r = i + is\cos(h)/\cos(\pi/3 - h) \\
&g = i + is(1 - \cos(h)/\cos(\pi/3 - h)) \\
&b = i - is \\
\\
&\text{For } h = 2\pi/3 \text{ to } 4\pi/3 : \\
&r = i - is \\
&g = i + is\cos(h - 2\pi/3)/\cos(\pi - h) \\
&b = i + is(1 - \cos(h - 2\pi/3)/\cos(\pi - h))
\end{aligned}
\tag{6.1}
$$

For $h = 4\pi/3$ to 2π :
$$r = i + is(1 - \cos(h - 4\pi/3)/\cos(5\pi/3 - h))$$
$$g = i - is$$
$$b = i + is\cos(h - 4\pi/3)/\cos(5\pi/3 - h).$$

The hue is an angle in the range of $0 \le h < 2\pi$ radians, the saturation is in the range of $0 \le s < 1$, and the intensity is an integer in the range of $0 \le i < 256$. An intensity of 0 is pure black, and an intensity of 255 is pure white. For our purposes, we take the saturation as 1.0 and the intensity as 128 and only vary the hue to produce a vivid rainbow of colors. The available colors for $i = 128$ are shown as a *color circle* in Fig. 6.1 for which the distance from the center is the saturation and the angle around the circle is the hue. Thus our colors will be chosen from those along the circumference of the circle.

Since we want adjacent colors to be very different while minimizing the repetition of similar colors, we calculate the hue from $h = 1.9416(N - 1)$ mod 2π where the constant 1.9416 is π divided by the *golden mean* $(1 + \sqrt{5})/2 = 1.6180...$ and N is the number of iterates for the orbit to escape. The golden mean (or *golden ratio*) is in a certain sense the 'most irrational'

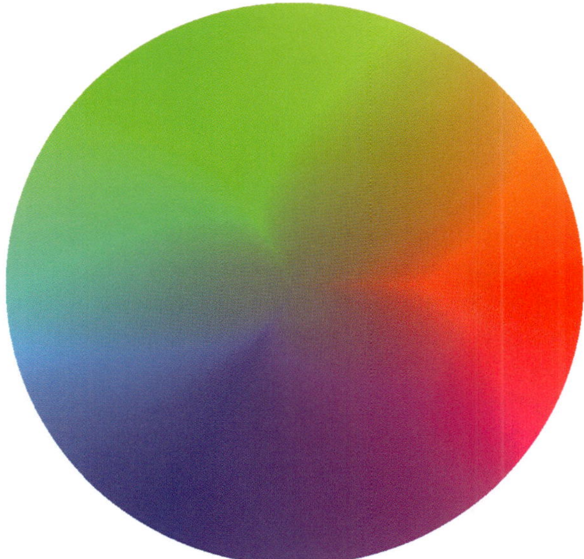

Fig. 6.1 The color circle for intensity $i = 128$ in which the saturation is represented by the radius (0 to 1) and the hue by the angle (0 to 2π).

of the irrational numbers [Livio (2002)], and thus it produces a sequence that is most unlike a periodic one.

6.3 Iterated Map Basins

You can apply the procedure to any of the attractors previously described. However, for variety we show here some new examples. The attractor is reduced in size in each dimension by a factor of three to allow room in the plot for its surrounding basin of attraction. Also realize that the basin plot is in a particular plane, taken here as $z = 0$ and that the attractor may have a dimension greater than two so that its projection can extend beyond the basin boundary in the chosen plane. It is as if you are looking at a three-dimensional object through a window, which does not ensure that the entire object is visible if you are some distance back from the window.

Figures 6.2 through 6.9 show some of the basins for iterated maps with an ampersand (&) appended to the code to indicate that it is an escape-time plot for the basin of attraction of the corresponding attractor. Note that symmetric attractors will generally have symmetric basins of attraction. You will probably agree that the basins of attraction are more elegant than the corresponding attractors. Most of the attractors that satisfy the necessary conditions for elegance (PF and CP) have elegant basins, at least for iterated maps, but many attractors that are not so elegant also have elegant basins.

6.4 ODE Basins

Attractors arising from systems of ordinary differential equations (ODEs) also have basins of attraction. However, chaotic attractors only occur when the dimension of the system, and hence the dimension of its basin, is three or greater. Thus it is necessary to show a cross section of the basin, here taken as $z = 0$.

Furthermore, the solution of a system of ordinary differential equations advances only very slowly with each iteration, and thus many iterations are required to test the boundedness of each initial condition. Those orbits that do escape tend to take many iterations to do so, and thus the escape-time contours are usually very closely packed, washing out the individual colors. A solution to this problem will be described later in the chapter. For the same reason, it is difficult to discern accurately the boundary of the basin in a reasonable amount of computation time. Fortunately, since our

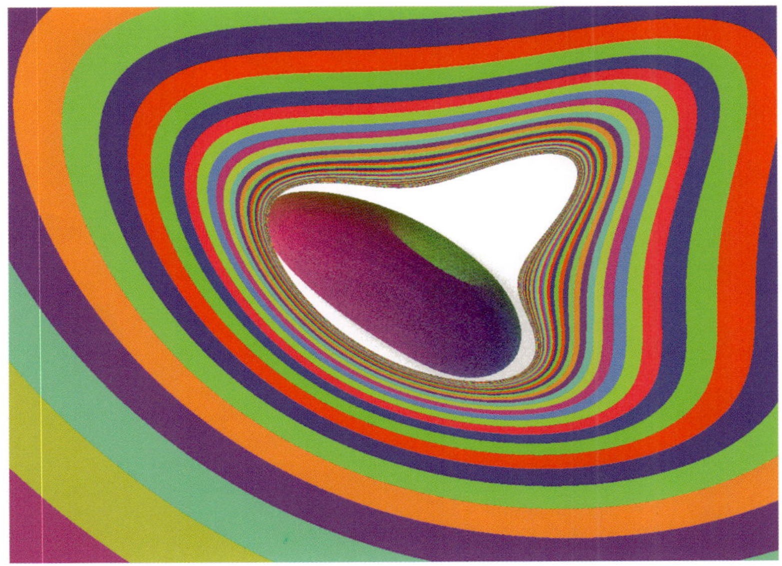

Fig. 6.2 `AMGYMNPEAPRLG&`.

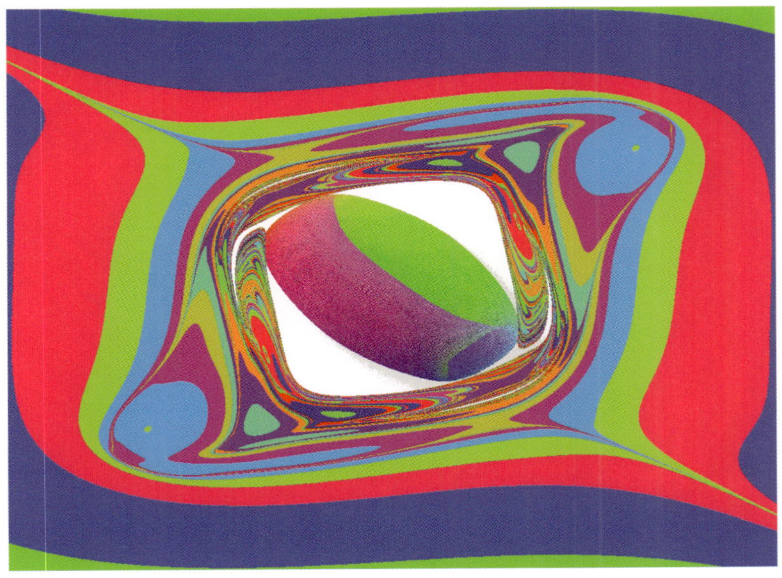

Fig. 6.3 `BERMZJHEEOLPN&`.

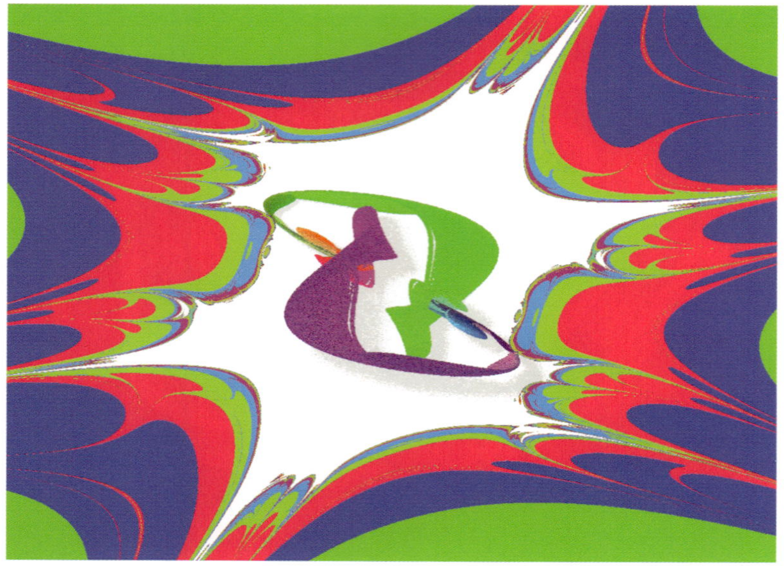

Fig. 6.4   CEKZPWSECKLTN&.

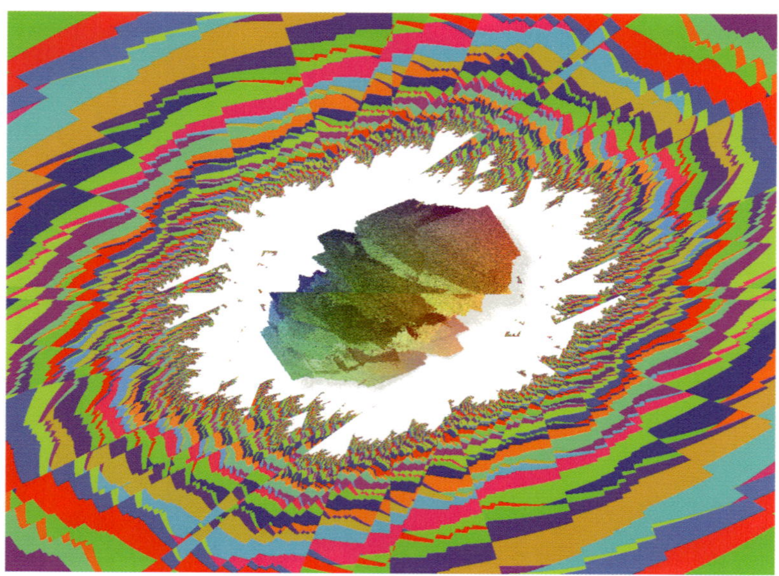

Fig. 6.5   DOZKYPDQKEJNF&.

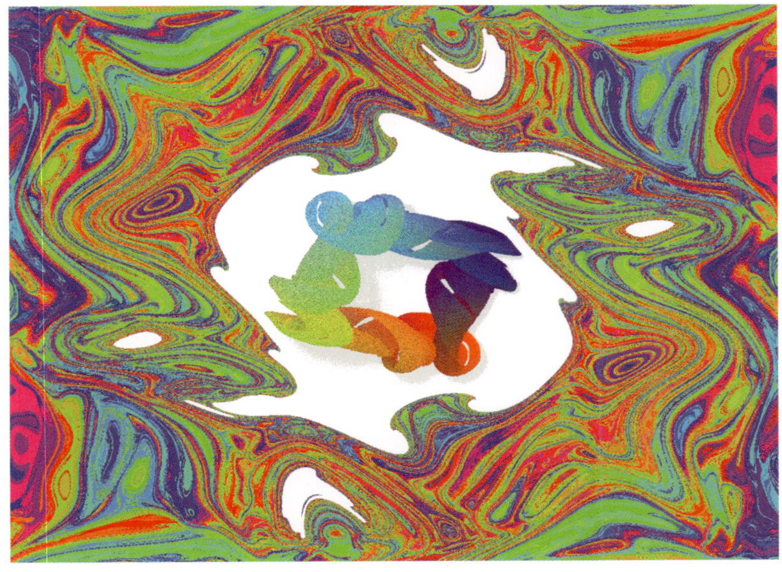

Fig. 6.6   EGOIBBRIEVDSQ&.

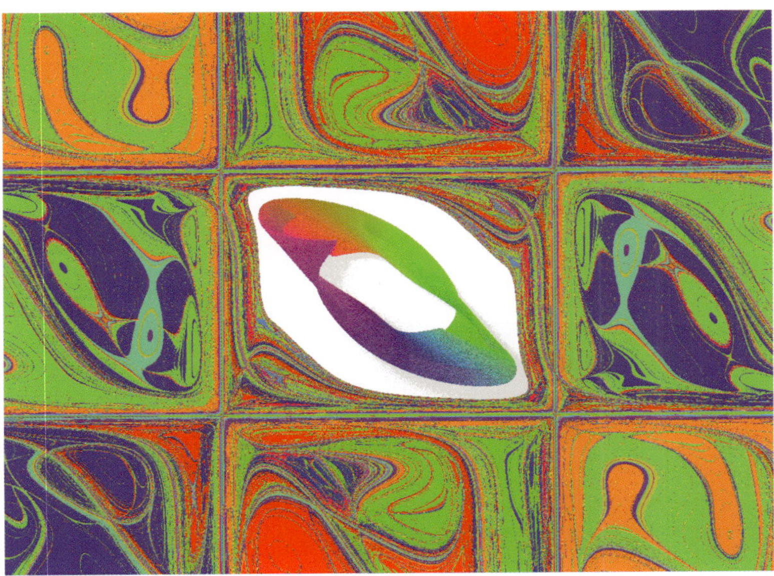

Fig. 6.7   FBMLOJTRQNBXT&.

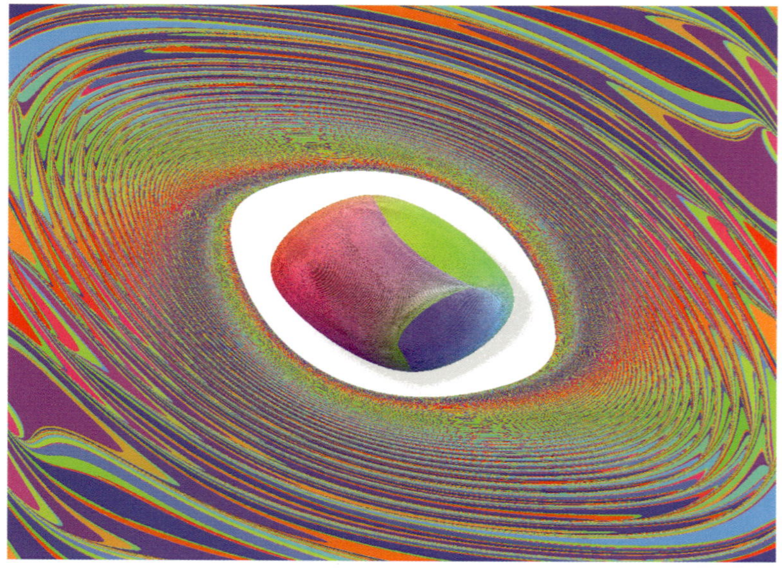

Fig. 6.8   GGRNHJAPYFACK&.

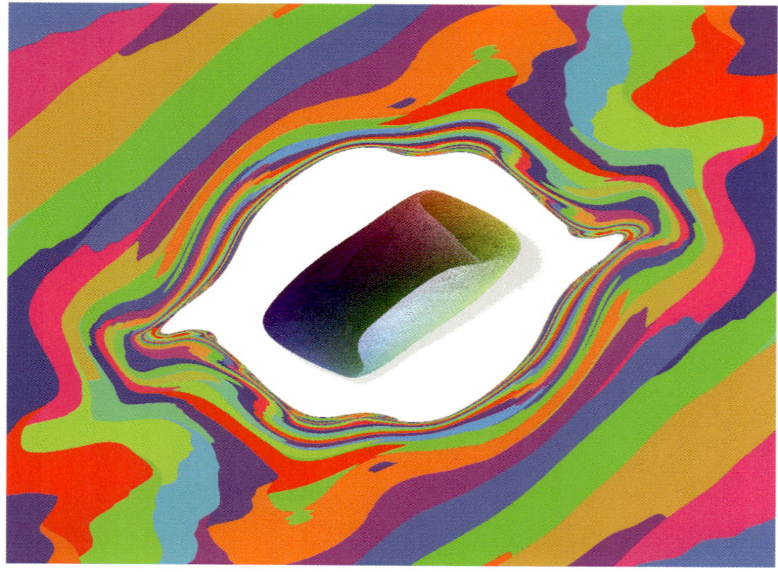

Fig. 6.9   HIXXPNSNQUBOW&.

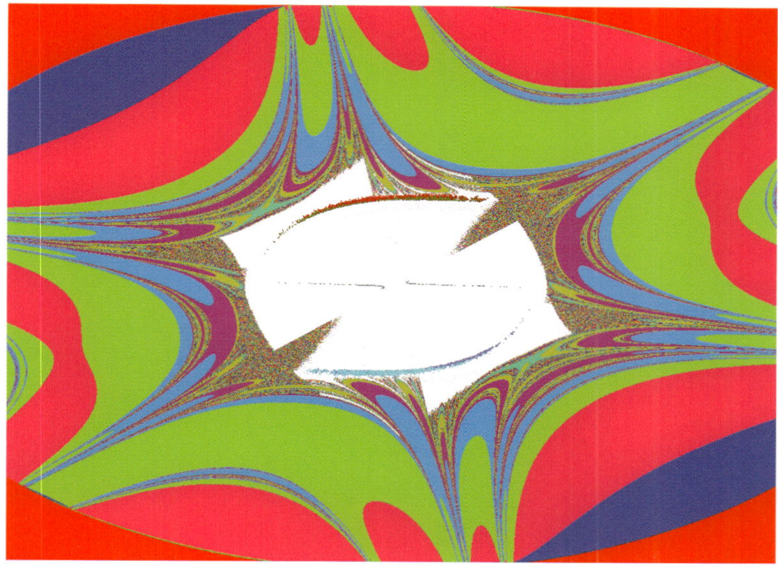

Fig. 6.10 `KEJZEXAORWWXI&`.

interest is mainly in making pretty pictures, these factors are not serious limitations.

The escape-time plots that arise from systems of ordinary differential equations are typically less elegant than those arising from iterated maps. However, Fig. 6.10 shows one example with interesting fractal structure.

6.5 Iterated Function System Basins

The attractors for iterated function systems also have basins of attraction. However, if the individual maps are linear, the basin will be the whole of space, and hence iterated orbits never escape. Thus they cannot provide escape-time contours. However, nonlinear iterated function systems typically do have finite basins of attraction, the only previous examples of which are the systems in Eqs. (5.5) and (5.6).

Another problem is that the random iteration algorithm causes initial conditions in adjacent pixels generally to have very different escape times. In fact, some sequences of random numbers give orbits that are in the basin, while others escape even for the same initial conditions. Thus the

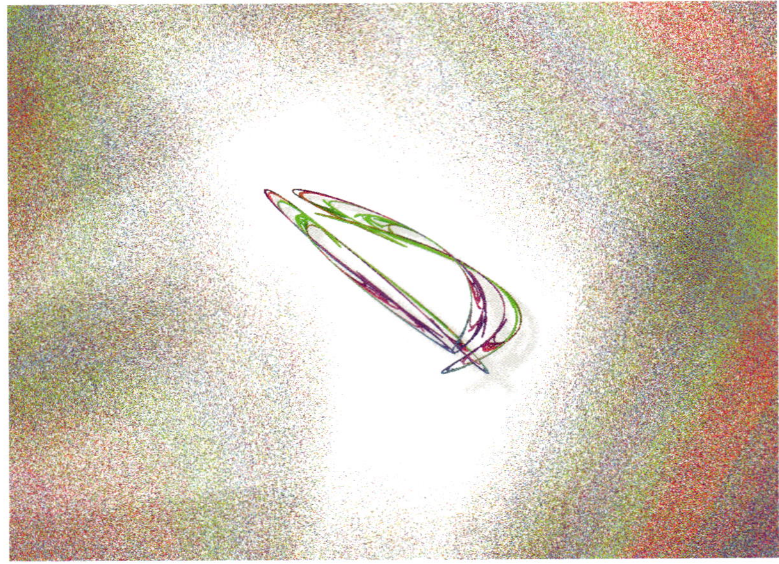

Fig. 6.11　`QSEPLHHXEJFXJ&`.

escape-time contours do not have distinct bands as in the previous examples. One such case shown in Fig. 6.11 gives an appearance of viewing the attractor through a fog.

A way to eliminate that problem is to reseed the random number generator with the same value for each initial condition, thus producing plots similar to those that use purely deterministic algorithms. The appearance of the resulting plot will depend on the seed used for the random numbers, and thus it provides an additional parameter for exploration. It is straightforward to do that, but rather than show examples, that will be left for you to explore.

6.6　General Escape-time Plots

All the preceding cases began by finding an elegant attractor and then calculating its basin of attraction. However, elegant escape-time plots can be produced even if the attractor just consists of a small number of isolated points or even if there is no attractor at all [Sprott and Pickover (1995)]. In such a case, there is no natural way to choose a size for the region of initial conditions to explore, and so we arbitrarily take $-1 < x < 1, -1 < y < 1$.

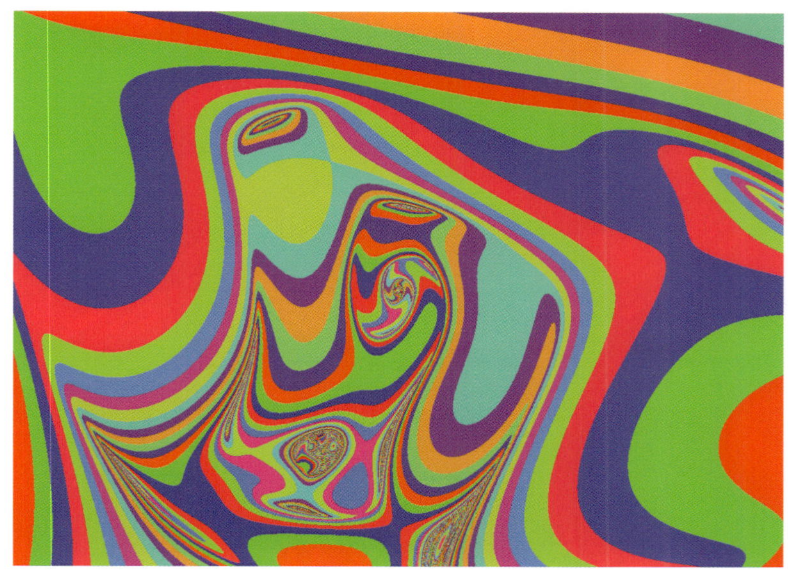

Fig. 6.12   ACLJIGIHCXLOK&.

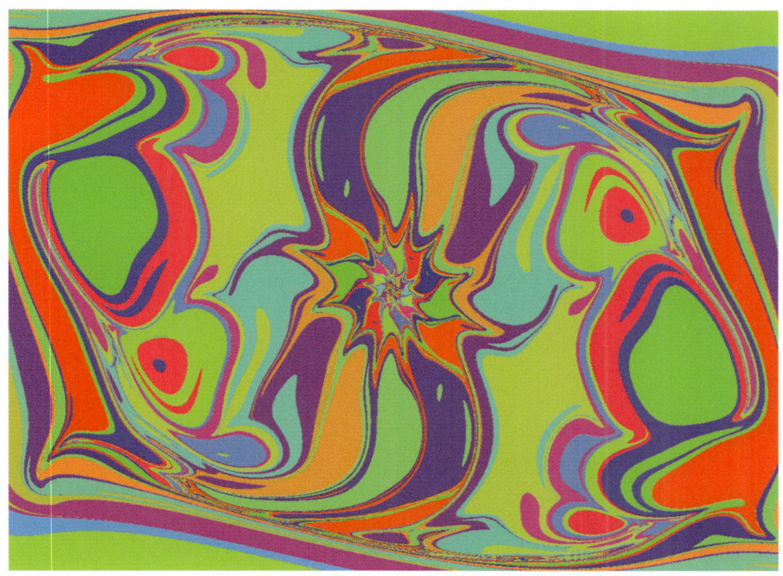

Fig. 6.13   BARLDLACWMKMB&.

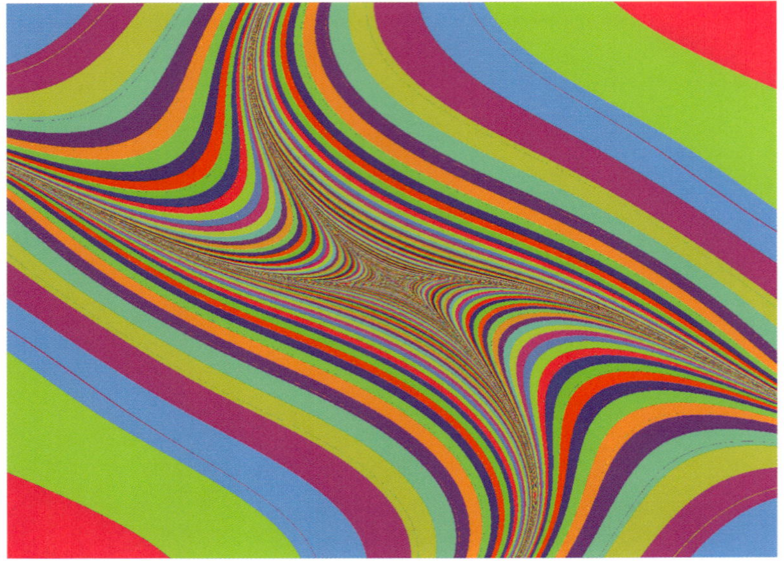

Fig. 6.14   CELIVBIQIDDVT&.

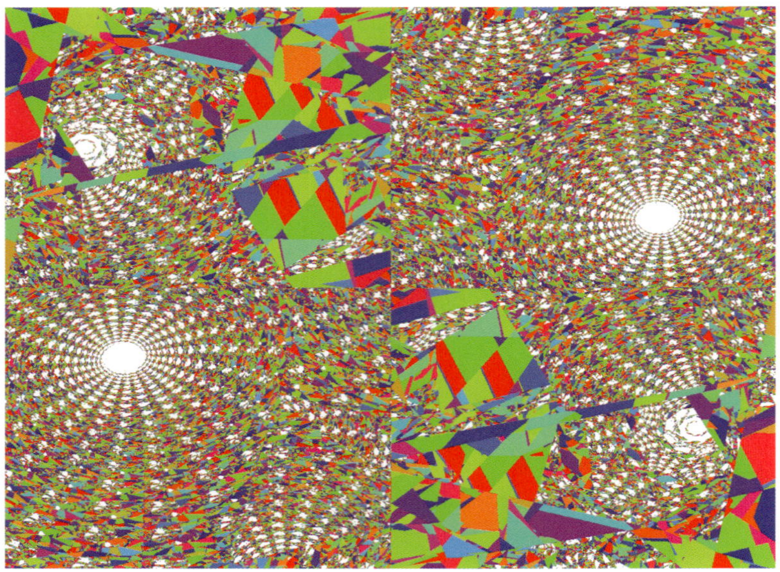

Fig. 6.15   DMXGTNYIWKTEN&.

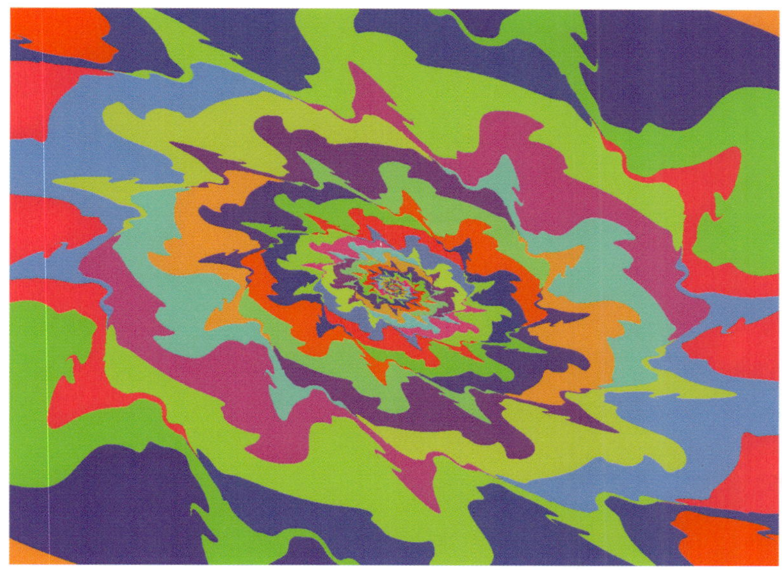

Fig. 6.16   EIQDBFMNFPMQU&.

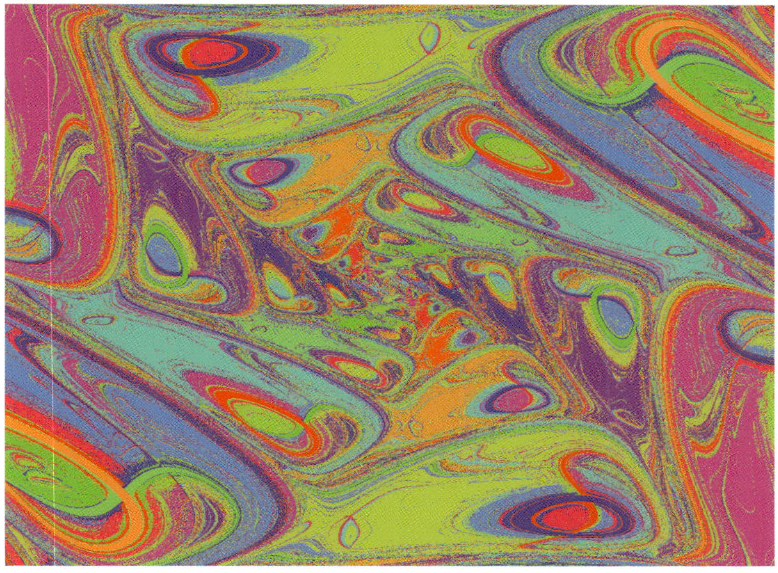

Fig. 6.17   FRQIDXPCJJEYM&.

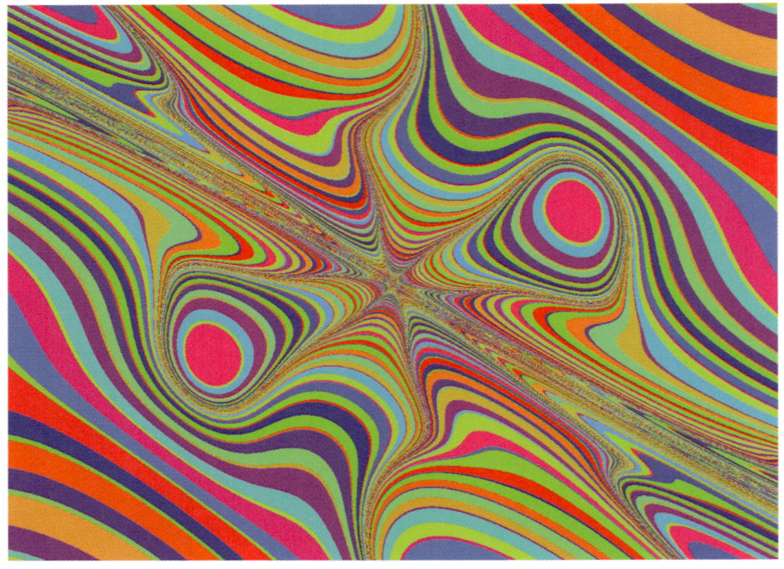

Fig. 6.18   GKDLZVDXWLMPN&.

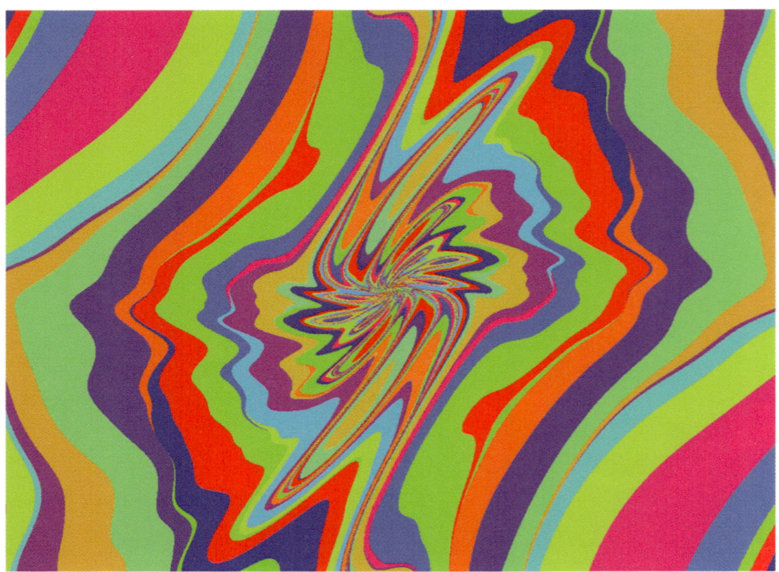

Fig. 6.19   HYCPJFKHWBZKH&.

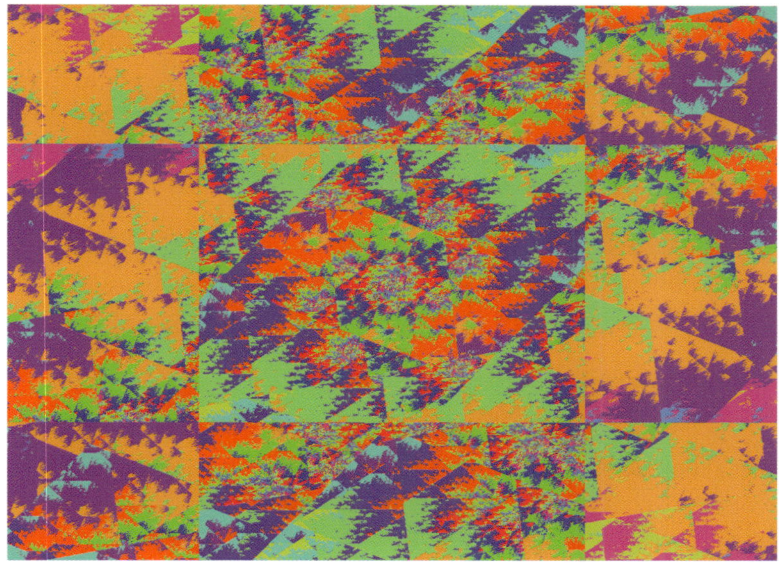

Fig. 6.20   IXOEVFQCKOLXL&.

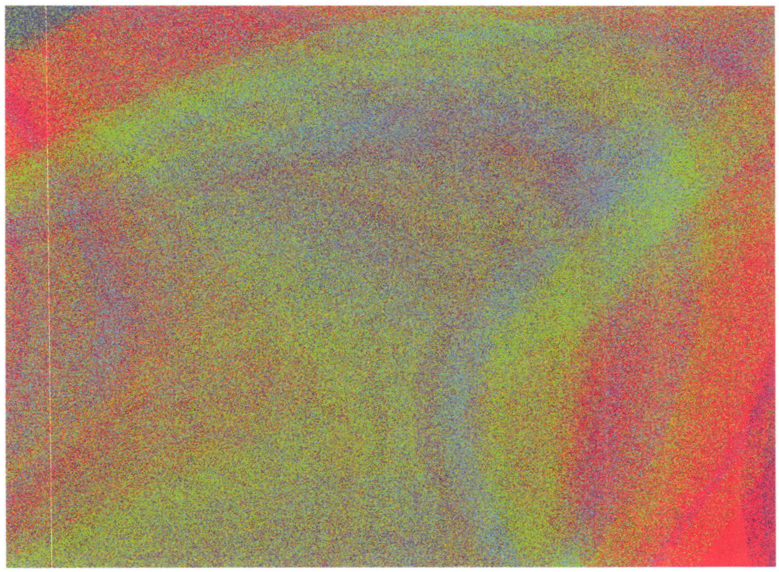

Fig. 6.21   QJVVYAUCPUATW&.

In addition, the criteria for choosing an elegant attractor are not useful, and some other means is required. For that purpose, we take an arbitrary initial condition in the chosen region and calculate the number of iterations required for it to escape. If that number is greater than 20 but less than 1000, the plot is likely to have interesting structure, at least in the vicinity of the chosen point. Cases that do not satisfy that criterion are discarded, and the test is relatively fast. Typical plots obtained in this way are shown in Figs. 6.12 to 6.21. Note that most of the plots have symmetry resulting from the symmetry of the governing equations. Symmetric attractors will generally have symmetric basins of attraction.

6.7   Generalized Julia Sets

Some of the most elegant fractals come from the simplest two-dimensional nonlinear iterated map, $z_{n+1} = z^2 + c$, where z and c are *complex numbers* given respectively by $z = x + iy$ and $c = a + ib$ with $i = \sqrt{-1}$. This system can be written more conveniently in terms of real variables as

$$\begin{aligned} x_{n+1} &= x^2 - y^2 + a \\ y_{n+1} &= 2xy + b. \end{aligned} \tag{6.2}$$

The set of initial conditions that remain bounded after infinitely many iterations is called the *Julia set* [Julia (1918)], or more properly, the *quadratic Julia set* since the nonlinearity is of the form of z^2. For almost every value of c ($c = 0$ and $c = -2$ are the only known exceptions), its basin boundary is a fractal, and an incredible variety of elegant escape-time plots can be produced by choosing appropriate values for the two parameters a and b [Peitgen and Richter (1986)].

Equation (6.2) can be generalized to include any two-dimensional map that satisfies the *Cauchy–Riemann equations* [Arfken (1985)] given by

$$\begin{aligned} \frac{\partial f}{\partial x} &= \frac{\partial g}{\partial y} \\ \frac{\partial f}{\partial y} &= -\frac{\partial g}{\partial x} \end{aligned} \tag{6.3}$$

where for Eq. (6.2), $f = x^2 - y^2 + a$ and $g = 2xy + b$. These equations ensure that the derivatives df/dz and dg/dz are independent of the direction in the complex plane along which the derivative is evaluated. However, maps that satisfy these conditions do not have chaotic attractors, at least when the nonlinearities are quadratic.

A system that satisfies Eq. (6.3) with twelve parameters is the two-dimensional quintic map

Case S:

$$x_{n+1} = a_1 + a_3 x - a_4 y$$
$$+ a_5 x^2 - 2a_6 xy - a_5 y^2$$
$$+ a_7 x^3 - 3a_8 x^2 y - 3a_7 xy^2 + a_8 y^3$$
$$+ a_9 x^4 - 4a_{10} x^3 y - 6a_9 x^2 y^2 + 4a_{10} xy^3 + a_9 y^4$$
$$+ a_{11} x^5 - 5a_{12} x^4 y - 10a_{11} x^3 y^2 + 10a_{12} x^2 y^3 + 5a_{11} xy^4 - a_{12} y^5$$

$$y_{n+1} = a_2 + a_4 x + a_3 y$$
$$+ a_6 x^2 + 2a_5 xy - a_6 y^2$$
$$+ a_8 x^3 + 3a_7 x^2 y - 3a_8 xy^2 - a_7 y^3$$
$$+ a_{10} x^4 + 4a_9 x^3 y - 6a_{10} x^2 y^2 - 4a_9 xy^3 + a_{10} y^4$$
$$+ a_{12} x^5 + 5a_{11} x^4 y - 10a_{12} x^3 y^2 - 10a_{11} x^2 y^3 + 5a_{12} xy^4 + a_{11} y^5.$$

$$(6.4)$$

Equation (6.2) is a special case of Eq. (6.4) with $a_1 = a, a_2 = b, a_5 = 1$, and all the other parameters equal to zero. Thus Eq. (6.4) is a *generalized Julia set*. Sample plots are shown in Figs. 6.22 to 6.25 with codes prefaced by S to indicate that they came from Eq. (6.4).

You may recognize the numerical values in the coefficients of Eq. (6.4) as *binomial coefficients* given by $n!/k!(n-k)!$, where n is the power of the variables, and k is the index of the term with $n \geq k \geq 0$. The coefficients can be arranged according to *Pascal's triangle* in which each value is the sum of the nearest two in the row above it:

$$
\begin{array}{ccccccccccccc}
 & & & & & & 1 & & & & & & \\
 & & & & & 1 & & 1 & & & & & \\
 & & & & 1 & & 2 & & 1 & & & & \\
 & & & 1 & & 3 & & 3 & & 1 & & & \\
 & & 1 & & 4 & & 6 & & 4 & & 1 & & \\
 & 1 & & 5 & & 10 & & 10 & & 5 & & 1 & \\
1 & & 6 & & 15 & & 20 & & 15 & & 6 & & 1 \\
1 & 7 & & 21 & & 35 & & 35 & & 21 & & 7 & 1.
\end{array}
$$

This should allow you to explore even higher-order polynomials, although you will probably find them to be not significantly more elegant than the examples given here but with a somewhat wider variety of patterns.

The attractors for these systems are a set of isolated points which do not satisfy the conditions for elegance. Thus you will be disappointed if you search this system for elegant attractors. It is ironic that the most elegant escape-time plots arise from the least elegant attractors whereas

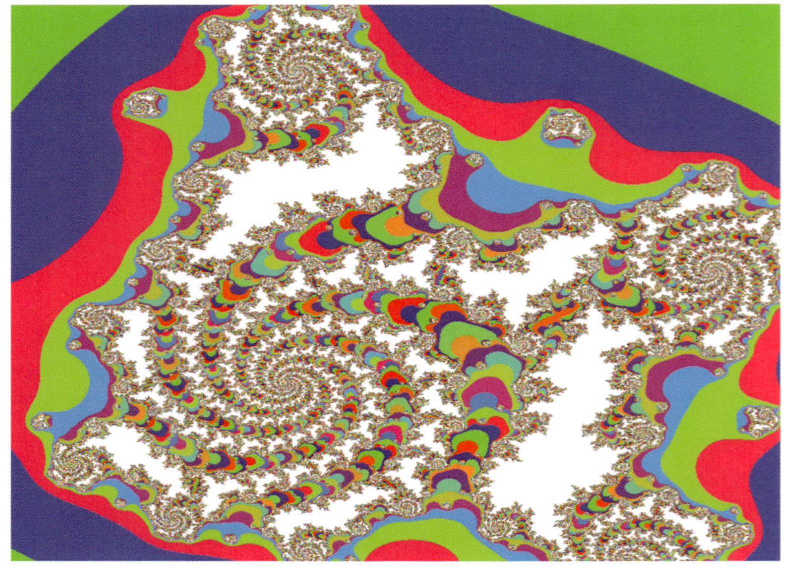

Fig. 6.22   SGIJQNKSQQHKG&.

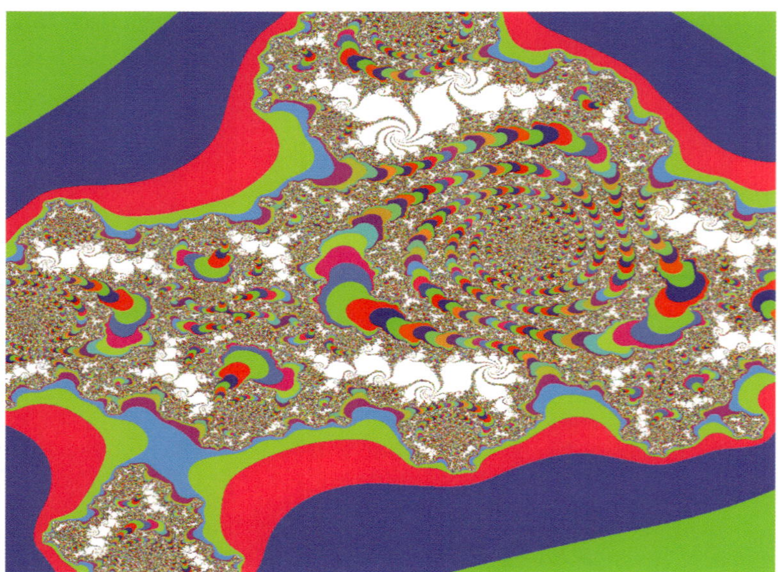

Fig. 6.23   SQRHHIHRJMMMT&.

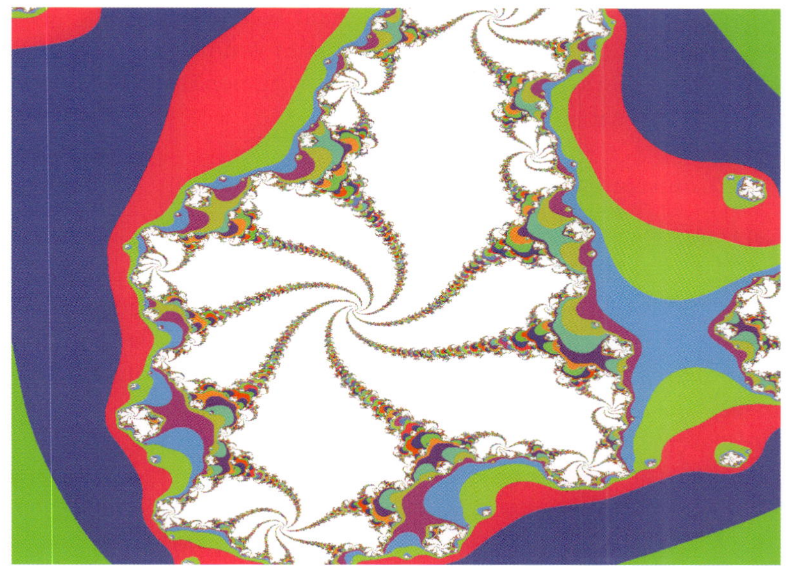

Fig. 6.24   SKMKVKJKROHSO&.

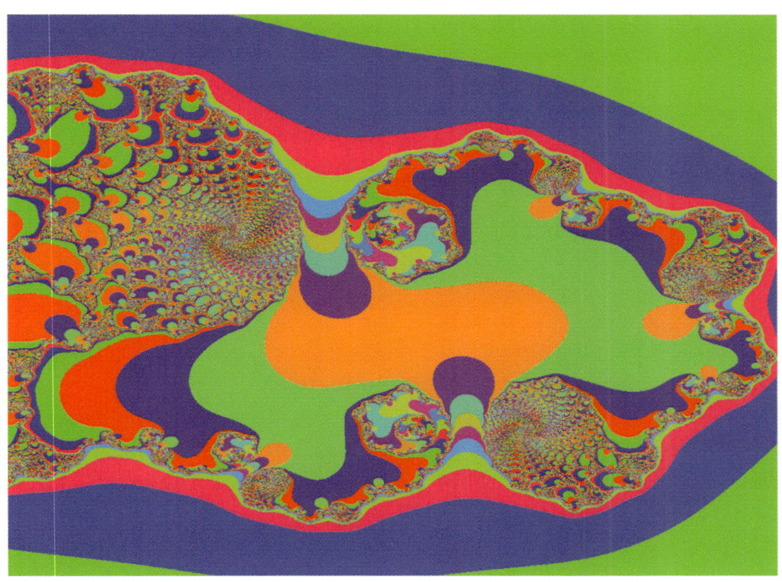

Fig. 6.25   SLNPKDDINSZPT&.

elegant attractors often have rather inelegant escape-time plots. The important criterion is that the Cauchy–Riemann condition in Eq. (6.3) be satisfied. You might wish to explore systems with nonlinearities other than polynomials that satisfy this condition.

6.8 Symmetric Julia Sets

You will note that the plots in Figs. 6.22 to 6.25 are not very symmetric. The reason is that the maps contain both odd and even powers of the variables. Since symmetry generally enhances elegance, you can constrain the parameters according to $a_1 = a_2 = a_5 = a_6 = a_9 = a_{10} = 0$, which gives a quintic map with only odd powers of the variables. Examples with that constraint are in Figs. 6.26 and 6.27.

Alternately, you can constrain the parameters according to $a_3 = a_4 = a_7 = a_8 = a_{11} = a_{12} = 0$ to give a quartic map with only even powers of the variables and whose escape-time plots are *antisymmetric* with examples in Figs. 6.28 and 6.29. Of the three types, asymmetric, symmetric, or antisymmetric, other things being equal, which do you find most elegant?

6.9 Elegant Equations

We could also ask what is the most elegant system of equations that gives rise to a fractal Julia set. Presumably such a system would be in the form of Eq. (6.2). There are precisely two such cases, one given by

$$\begin{aligned} x_{n+1} &= x^2 - y^2 - 1 \\ y_{n+1} &= 2xy \end{aligned} \tag{6.5}$$

with an escape-time plot as in Fig. 6.30, and the other given by

$$\begin{aligned} x_{n+1} &= x^2 - y^2 \\ y_{n+1} &= 2xy \pm 1 \end{aligned} \tag{6.6}$$

with an escape-time plot as in Fig. 6.31.

These two cases can be written more compactly in terms of complex variables as $z_{n+1} = z^2 - 1$ and $z_{n+1} = z^2 \pm i$, respectively. The case with $c = -1$ from Eq. (6.5) in Fig. 6.30 contains infinitely many slightly distorted *Mandelbrot sets*. The *Mandelbrot set* is the set of all values of c for which the Julia set with initial conditions $z = 0$ is connected. The case with $c = +i$ from Eq. (6.6) in Fig. 6.31 is called a 'dendrite,' and it lies precisely on the boundary of the Mandelbrot set. It is remarkable that such a simple equation gives such elegant plots.

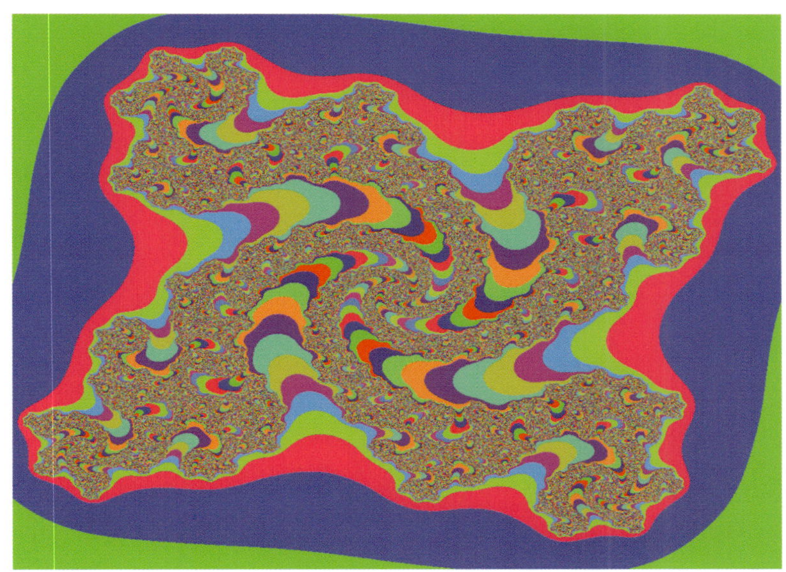

Fig. 6.26   SMMKXMMPMMMCW&.

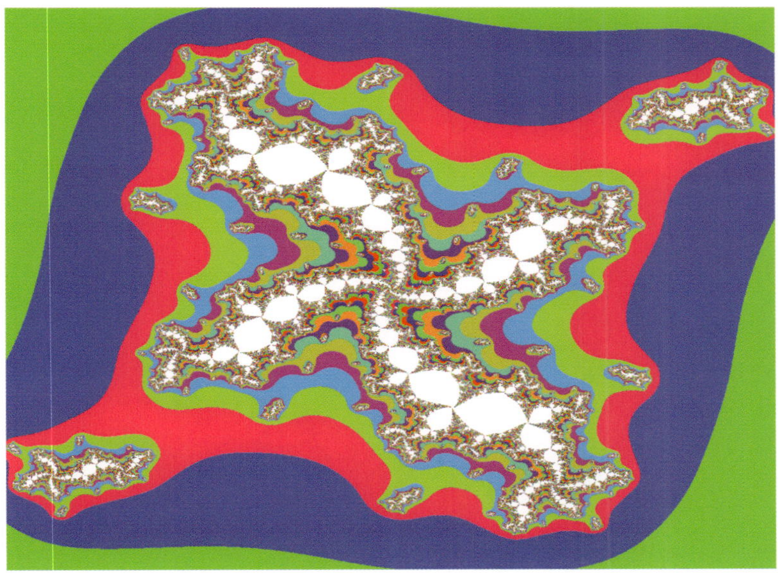

Fig. 6.27   SMMNAMMGGMMXC&.

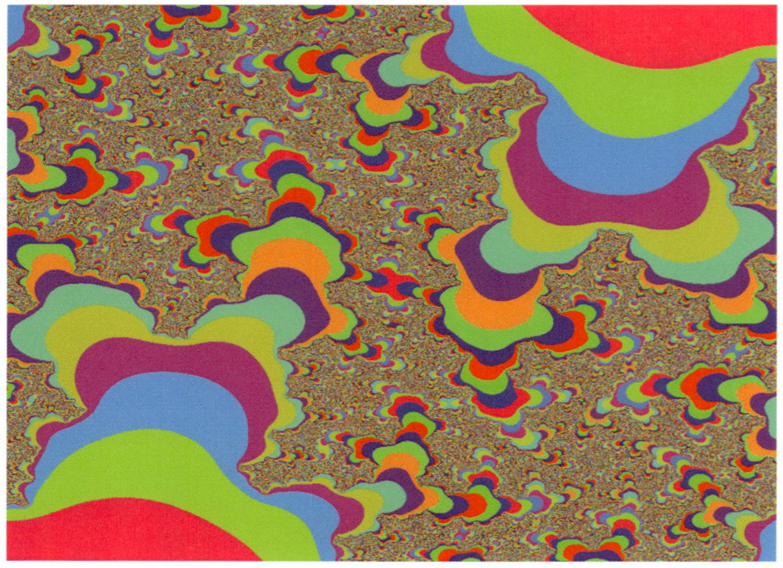

Fig. 6.28   SVEMMKHMMNNNMM&.

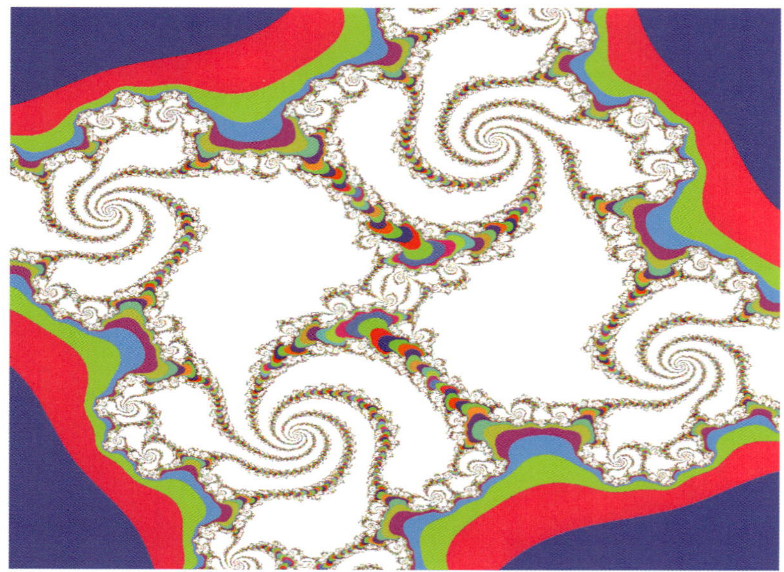

Fig. 6.29   SEQMMKMMMSPMM&.

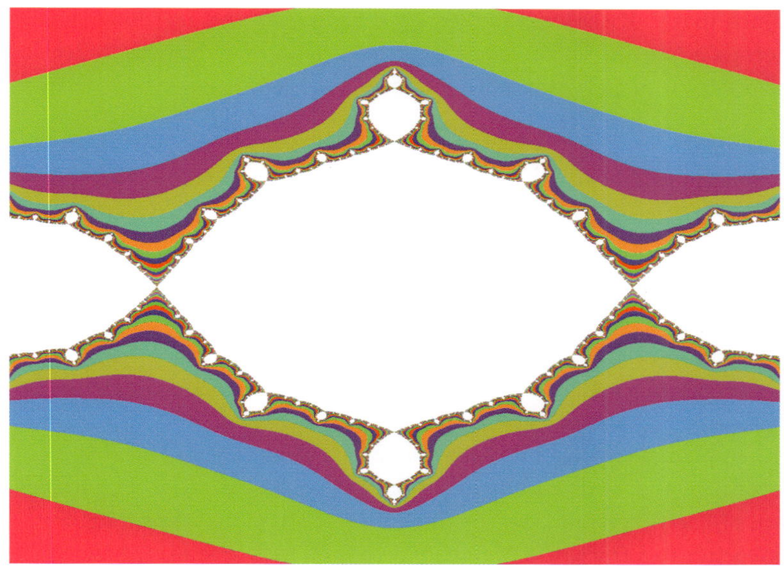

Fig. 6.30   SCMMMWMMMMMMM&.

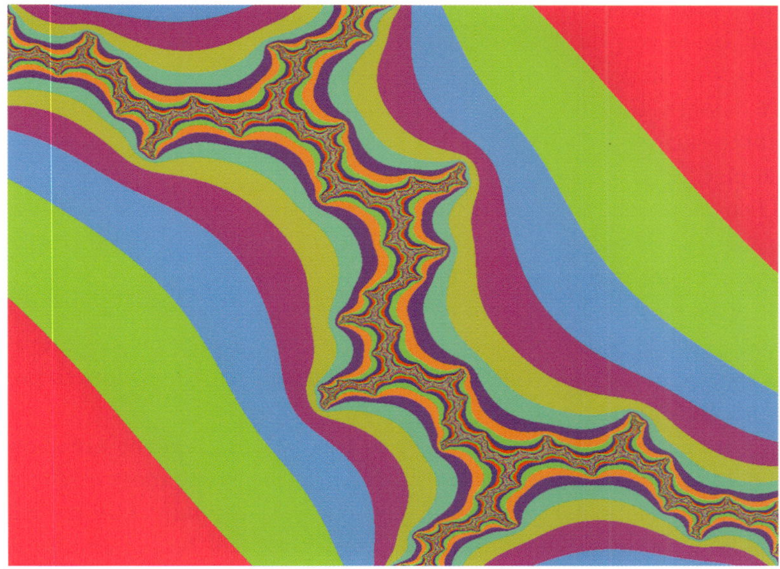

Fig. 6.31   SMWMMWMMMMMMM&.

Since many of these examples have right/left and up/down symmetry, numerous copies of them could be placed side-by-side with reflections to make a very unusual fractal wallpaper.

6.10 Alternate Coloring Scheme

While the coloring scheme employed in the previous examples is simple to implement and works especially well if all you have is black and white, the resulting plots are somewhat inelegant. The sharp boundary between colors and the rapid cycling of colors near the basin boundary may disturb you. The former problem can be corrected using interpolation, and the latter can be corrected using the logarithm of the number of iterations required for the orbit to escape.

To implement interpolation, calculate the square of the distance from the origin $r^2 = x^2 + y^2$ for the iteration N that first exceeds $r = 100$ and the square of the corresponding distance for the previous iterate $r_0^2 = x_0^2 + y_0^2$. Then calculate a fraction δN in the range of 0 to 1 to subtract from N using the formula $\delta N = \log(r^2/10^4)/\log(r^2/r_0^2)$. The interpolation is logarithmic rather than linear because r typically grows exponentially at large distances from the origin.

Note that since δN depends on the ratio of two logarithms, any convenient bases can be used as long as they are identical. Since $\log(r^2) = 2\log(r)$, you might be tempted to simplify the above interpolation formula to $\delta N = \log(r/100)/\log(r/r_0)$, but that actually slows the calculation since it unnecessarily requires taking the square root of r^2 and r_0^2 at each pixel of the plot.

Then choose the color using Eq. (6.1) with a hue given by $h = 3\pi \log(N - \delta N)/\log(N_{bail})$, where N_{bail} is the maximum number of iterations before concluding that the orbit will not escape (typically $N_{bail} = 1000$). The factor of 3π is chosen to ensure that the hue cycles through the entire color circle (2π) while ending with a color that contrasts with the starting color. You can explore factors other than 3 and different starting colors to provide more variety in the plots.

Some sample images produced with this coloring scheme are shown in Figs. 6.32 to 6.38. Probably you will agree that they are more elegant than the previous examples, and they likely represent the most elegant fractals in this book, although with an appearance that will be familiar to you if you have seen previous examples of fractal artwork [Peitgen and Richter (1986)].

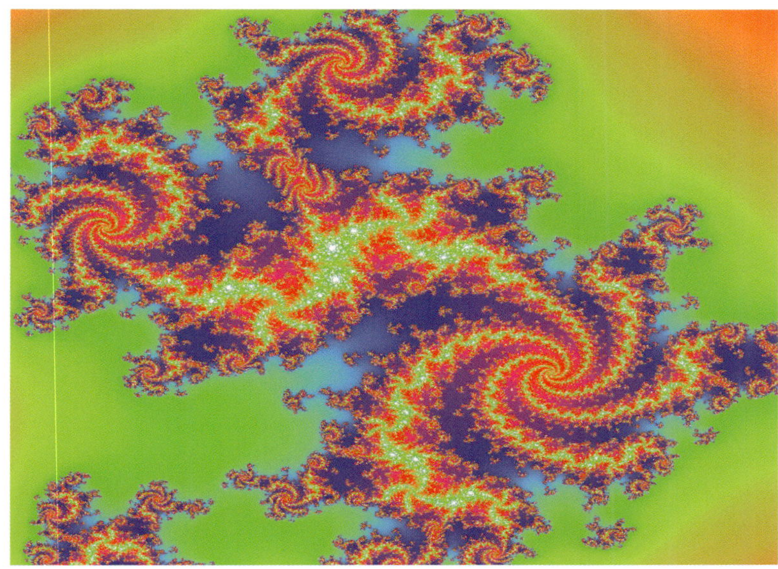

Fig. 6.32 SRMNKSIPLLIGO&.

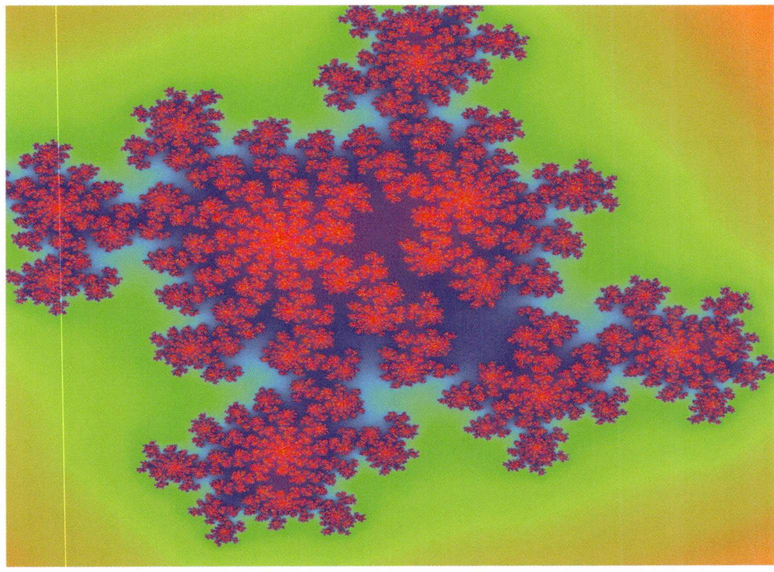

Fig. 6.33 SEPMOJLTPTUNL&.

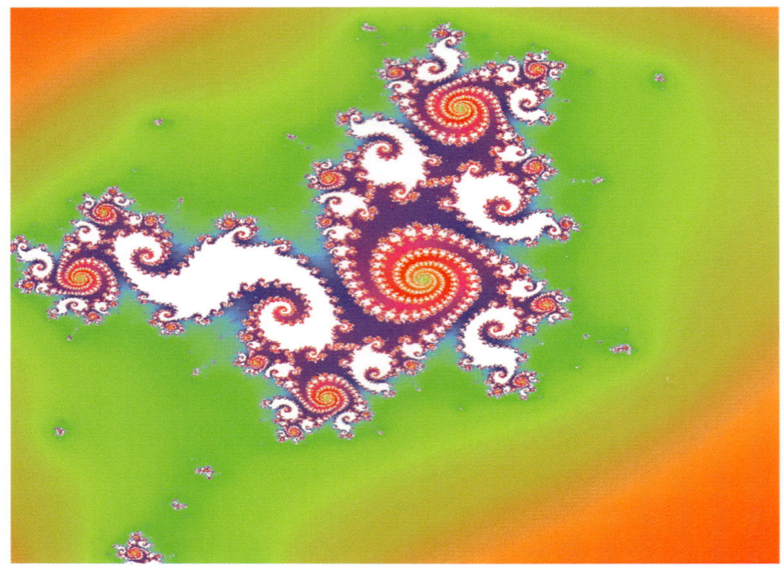

Fig. 6.34   SNNFFPBZDNKOW&.

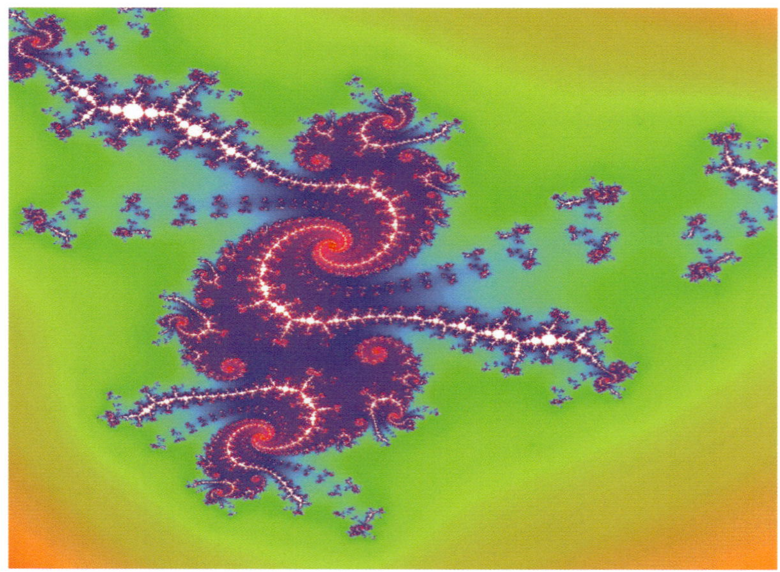

Fig. 6.35   SJPDOISNDUJJM&.

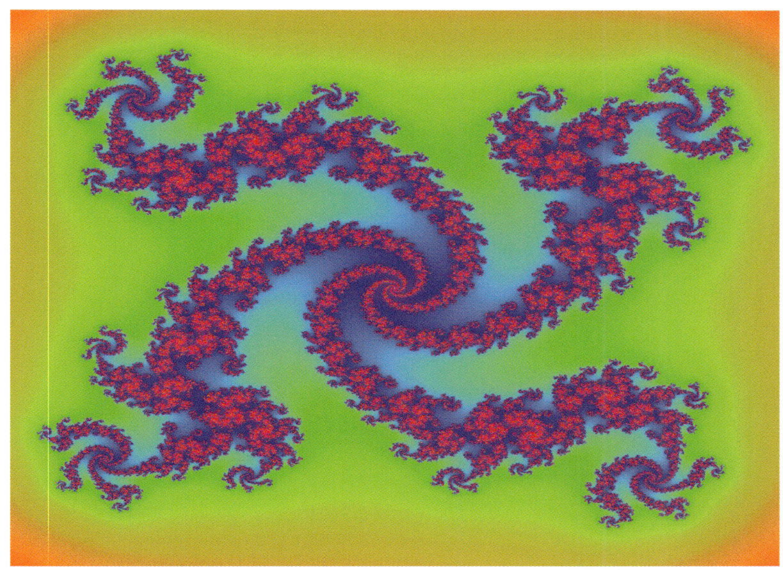

Fig. 6.36 SMMOBMMLOMMRA&.

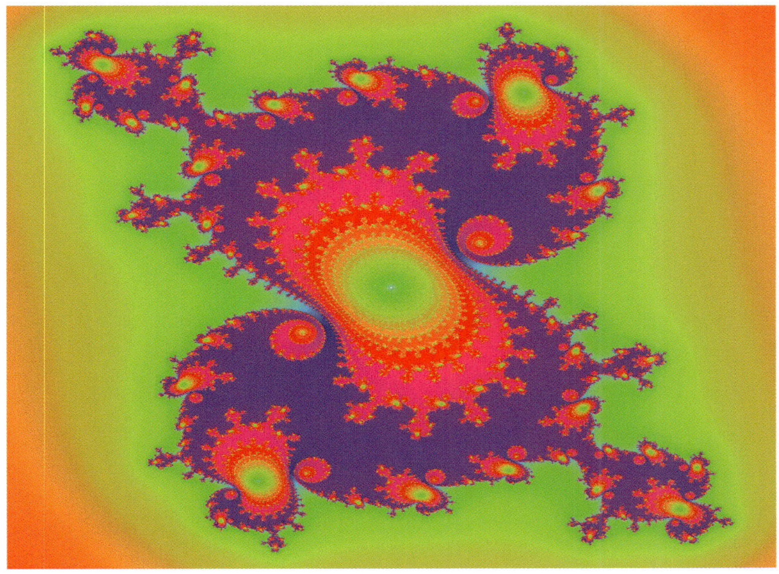

Fig. 6.37 SMMWLMMXQMMXD&.

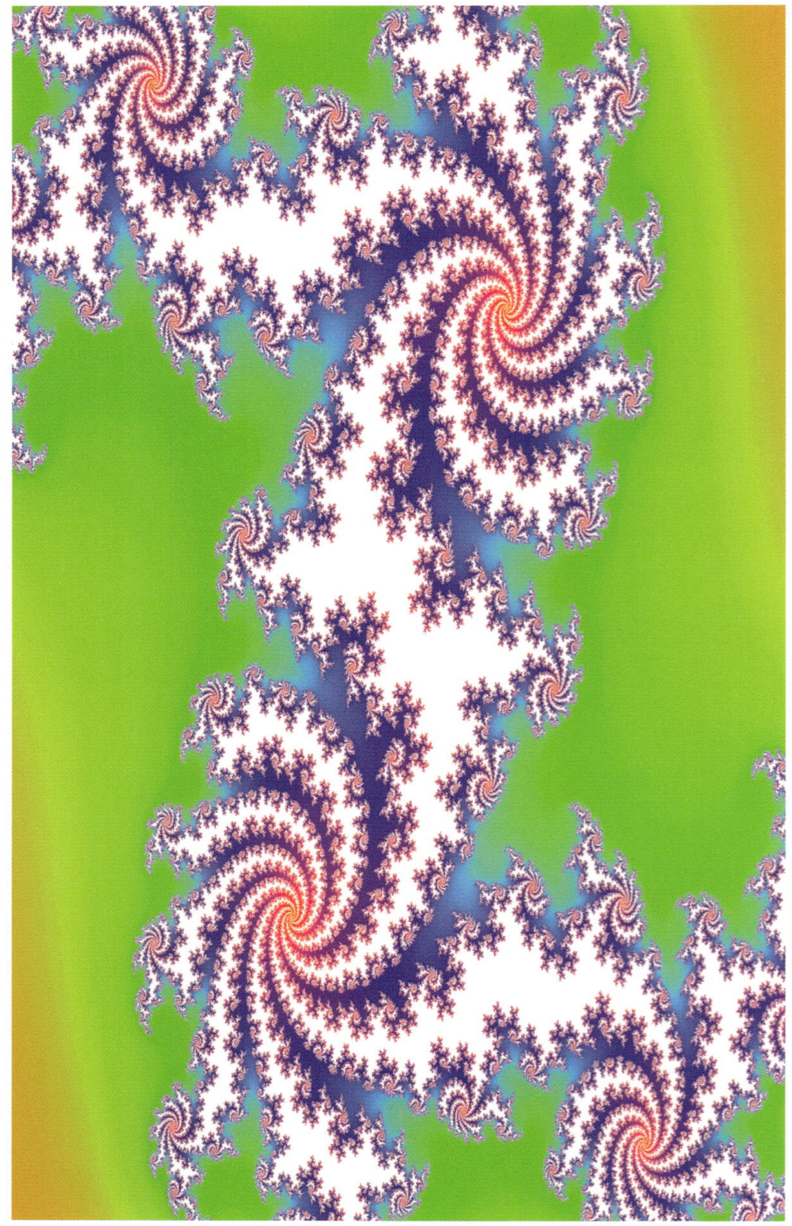

Fig. 6.38　SHIMMGKMMHQMM&.

Chapter 7

Spatiotemporal Systems

All the previous examples involved systems in which single orbits were followed in time as they visited successive points in space. Here we consider systems with many points in space that evolve simultaneously according to their state and the state of their near neighbors. Such systems are more computationally intensive, but they represent better models of real-world systems such as weather patterns and landscapes, even if the resulting images are generally less elegant than the previous ones.

7.1　Elementary One-dimensional Cellular Automata

Perhaps the simplest example of a spatiotemporal system is a *cellular automaton* in which each site (or *cell*) in space has a finite number of states, which in the simplest case is only two and can be called *dead* and *alive*, or *off* and *on*, or *white* and *black*, or 0 and 1, or *Democrats* and *Republicans*, or any other pair of opposing quantities you can imagine. Such systems were introduced by Stanisław Ulam (1952) and John von Neumann (1966) at Los Alamos during the 1940s and have been extensively studied by Stephen Wolfram (2002) and many others.

The simplest cellular automaton has the sites arranged along a line in one dimension, usually with periodic boundary conditions (the right end of the line is assumed to be adjacent to the left end, or if you prefer, imagine the line wrapped into a circle so that it has no ends). Time advances in discrete steps, and all the points along the line are updated according to a rule that depends on the value at each site and some number of its near neighbors. The rule is customarily taken to be the same for all the sites (everyone plays by the same rules).

Usually the sites are updated synchronously, meaning that the whole

153

line is updated at the same time, which requires that the previous line along with the current line be stored in memory, and that is the method used here. Alternately, the sites can be updated sequentially (one at a time) or stochastically (randomly), which are easier to program but lead to slightly different results.

In the simplest nontrivial case, the rule involves only the two nearest neighbors to each site, the one to the left and the one to the right. Since there are eight possible configurations of a site and its two nearest neighbors (000, 001, 010, 011, 100, 101, 110, and 111), each of which can give rise to two possible results, there are $2^8 = 256$ different rules to determine the next value for each site along the line. These elementary examples have been indexed as a number from 0 to 255 and catalogued by Wolfram (1983).

Most of the rules lead to uninteresting behavior such as all the sites becoming identical or alternating periodically in space or time, but some give rise to quite remarkable behaviors, two examples of which are in Figs. 7.1 (rule 30) and 7.2 (rule 126). Wolfram (2002) classifies cellular automata into four classes in order of complexity: Class 1 evolves into a stable homogeneous state. Class 2 evolves into stable or oscillating structures. Class 3 appears to be chaotic with aperiodic oscillations. Class 4 exhibits complex interacting structures that are capable of universal computation.

As is customary, these figures display the lines horizontally with time advancing downward so that successive generations can be observed simultaneously. When the plot reaches the bottom (after 900 generations), it begins again at the top, overwriting the previous plot until a total of $10 \times 900 = 9000$ generations of the 1200 points along the line have been calculated. However, in this case, the calculation is stopped when it first reaches the bottom to show only the first 900 generations. Cells with a value of zero are plotted as white, and those with a value of 1 are plotted in one of two colors that are determined by the previous value of each cell. For each case, the initial condition is taken as a single cell equal to 1 at the center top of the plot, with all other cells equal to zero.

Rule 30 in Fig. 7.1, which looks rather bland, is of interest because it is chaotic [Cattaneo *et al.* (2000)]. The sequence of bits that occur down the center axis passes most tests for randomness [Wolfram (1986)] and is used as a pseudo-random number generator in the *Wolfram Mathematica* symbolic computation program [Wolfram (2017)].

Rule 126 in Fig. 7.2 consists of *Sierpiński triangles* [Sierpiński (1916)], which is a classic fractal that has been widely studied for over a century. It has a fractal dimension of $D0 = \log(3)/\log(2) \approx 1.585$. This is only one of

Fig. 7.1   Elementary cellular automaton from rule 30 ($PF = 0.494, CP = 0.030$).

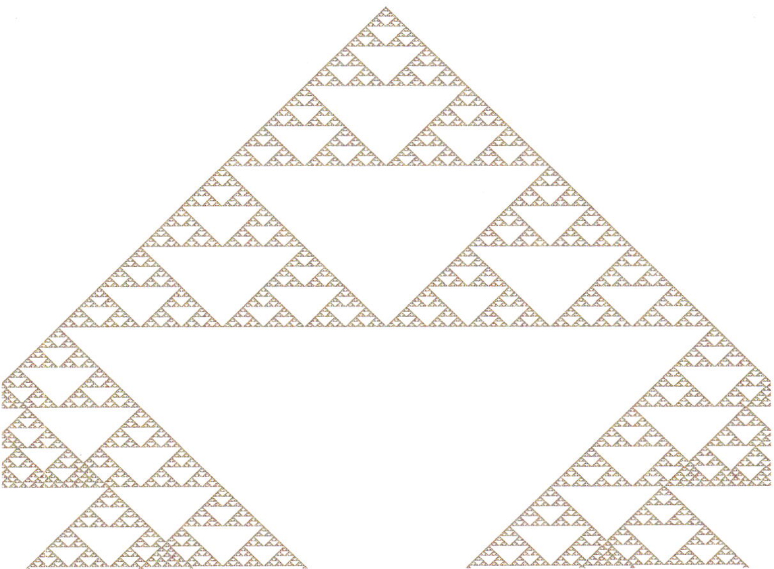

Fig. 7.2   Elementary cellular automaton from rule 126 ($PF = 0.037, CP = 0.182$).

about a dozen different ways to produce this particular classic fractal. It is a triangular version of the Sierpiński carpet in Fig. 1.4.

7.2   General One-dimensional Cellular Automata

The previous classic cellular automata are instructive, but they are far too limiting for our purposes of generating an enormous variety of fractal patterns since there are only 256 possible rules. One simple generalization is to allow each site to depend on its previous state and the state its four (rather than two) nearest neighbors. The corresponding patterns then consist of five-bit numbers, which can take on $2^5 = 32$ values, giving $2^{32} = 4\,294\,967\,296$ possible rules.

Following the naming convention for previous examples in this book, we preface the code with a T to denote a one-dimensional, two-state cellular automaton and use the next eight characters to specify the particular rule. We use an A as the first character of the rule to indicate that it is a general four-neighbor rule and the next seven characters to specify the rule according to

> Case **TA** to **TH**:
>
> | Parameter | Used for |
> |---|---|
> | a_1 | Set to -1.2 (**A**) to -0.5 (**H**) |
> | a_2 to a_8 | $rule = \sum\limits_{i=0}^{6} 26^i(12 + 10a_{i+2}) \mod 4\,294\,967\,296$ |
> | a_9 | Initial conditions (see text) |
> | a_{10} | Color (see text) |
> | a_{11} | Replace color with white if $a_{11} \leq 0$ |
> | a_{12} | Not used, set to 0 (**M**). |

Thus the seven characters in the code (following the **TA**) represent a seven-digit base-26 random number from which the rule is generated. (The base of 26 is also called the *radix* of the number system.) Since there are $26^7 = 8\,031\,810\,176$ possible values of *rule*, the *modulo* (mod) function is used to restrict the values to less than 2^{32} while introducing some unnecessary but harmless redundancy in the notation (some identical rules can be expressed by two different codes).

If we retain twelve characters in the code for each case, that allows four additional parameters for other purposes. We will use a_9 to specify the initial conditions on the first row. If $a_9 < 0$, the initial conditions will be spatially periodic with $10|a_9|$ cycles. If $a_9 = 0$, there will be a single one at the center of the line with all other sites zero, and if $a_9 > 0$, the initial conditions will be a random sequence of zeros and ones. In particular, a one will be used if $ran < a_9/1.4$ and a zero otherwise, where ran is a uniform random number in the range of 0 to 1. This allows control of the initial fraction of ones from about 7% (if $a_9 = 0.1$) to about 93% (if $a_9 = 1.3$).

In addition, rather than using only two colors determined by the previous value of each site, we will choose among 32 colors as given by the five-bit pattern of the previous neighborhood of each point using Eq. (6.1) with an integer color offset of $12 + 10a_{10}$ to give more variety of colors. The parameter a_{11} will be used to control whether sites with a value of zero are plotted in color (if $a_{11} > 0$) or otherwise in white. The last parameter a_{12} is not used and is set to zero (M).

Even with these restrictions, there are still over 3×10^{12} possible patterns as determined by the randomly chosen parameters. Figures 7.3 and 7.4 show two examples obtained in this way in which you are seeing generations 8101 to 9000. In these and the following images, the contrast has been adjusted using a photo editing program to improve visibility of the spatial structure.

We can also constrain the rule so that it has left/right symmetry with respect to each site. This is not to say that the resulting pattern will have such a symmetry, but only that the two sides of each site contribute in the same way to its next value. There are still five configurations of each site and its two nearest neighbors, and thus there are $2^5 = 32$ different configurations, but rather than $2^{32} = 4\,294\,967\,296$ different rules, there are now only $2^{16} = 65536$ rules since the last 16 bits of the rule are the mirror image of the first 16 bits. Patterns produced by such a symmetric rule will be denoted by the code TB, two examples of which are in Figs. 7.5 and 7.6.

Another possibility is to constrain the rule so that it is right/left *antisymmetric*. That is to say that the last 16 bits of the rule will be opposite to the first 16 bits (zeros and ones are exchanged). Patterns produced by such an antisymmetric rule will be denoted by the code TC, two examples of which are in Figs. 7.7 and 7.8.

A different type of rule involves only the total (or equivalently, the average) of the cells in the neighborhood of each site. This considerably reduces the size of the rule space from 2^{2^m} to 2^{2m+2}. where m is the size of the neighborhood, which allows you to explore larger neighborhoods for a given rule space. For example, consider the seven neighbors on each side ($m = 15$). If each site (including the one at the center) can be a zero or a one, there are sixteen possible totals ranging from 0 (all zeros) to 15 (all ones). Since the center site can be either a zero or a one, there are

 Elegant Fractals

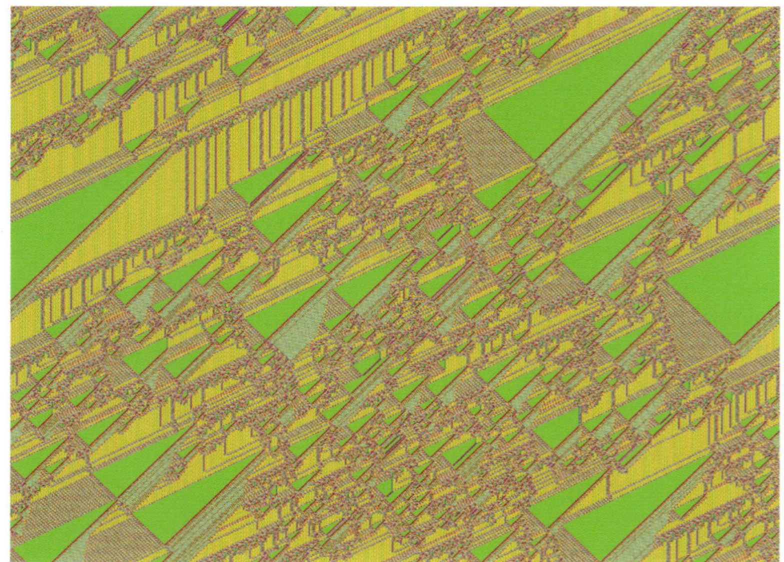

Fig. 7.3 **TAOVZWMVVUMTM** $(PF = 0.465, CP = 0.039)$.

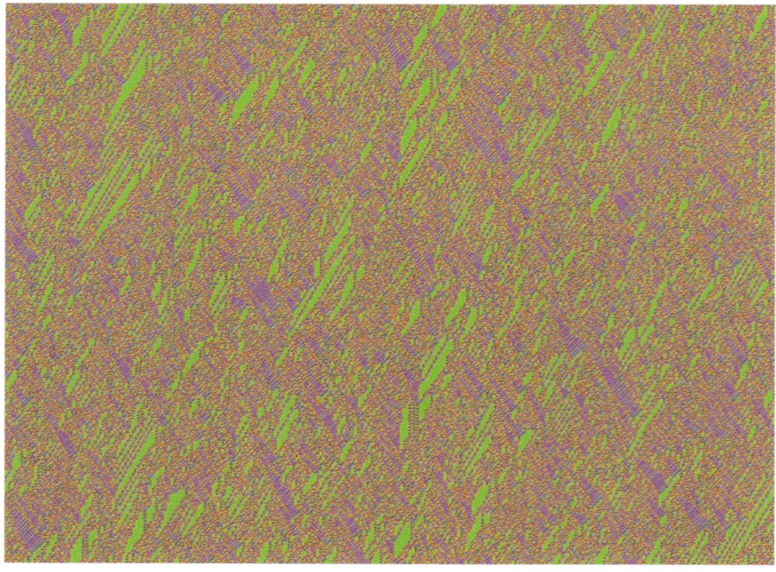

Fig. 7.4 **TAZAYZQYSHGWM** $(PF = 0.570, CP = 0.218)$.

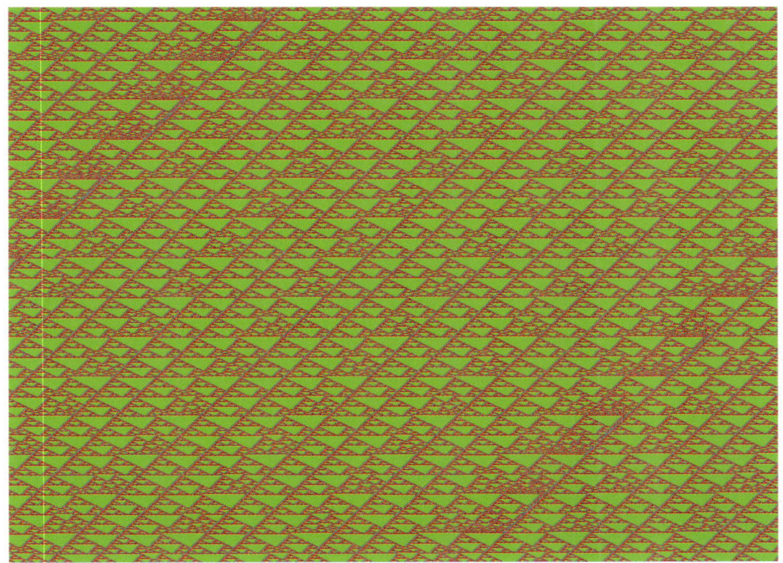

Fig. 7.5   **TBKZTNDPEDMTM** $(PF = 0.376, CP = 0.306)$.

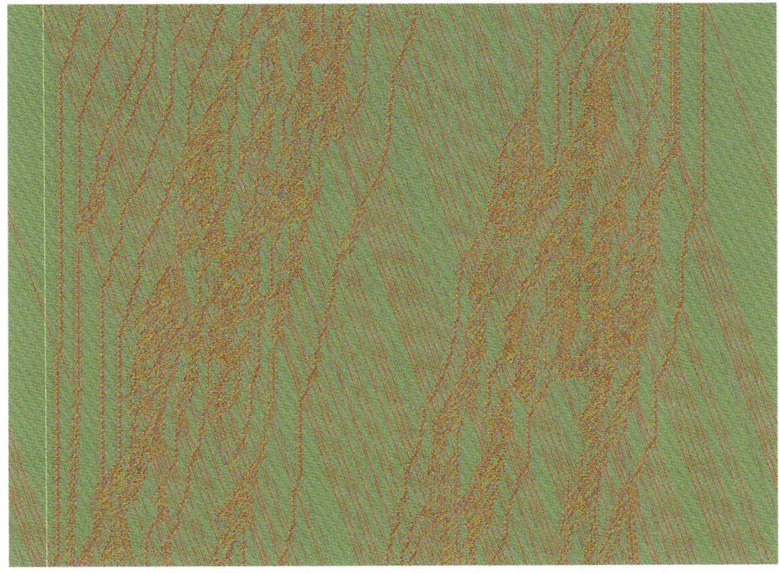

Fig. 7.6   **TBDYCXDODGMUM** $(PF = 0.394, CP = 0.032)$.

 Elegant Fractals

Fig. 7.7   TCGKYWNNGLKMM ($PF = 0.528, CP = 0.342$).

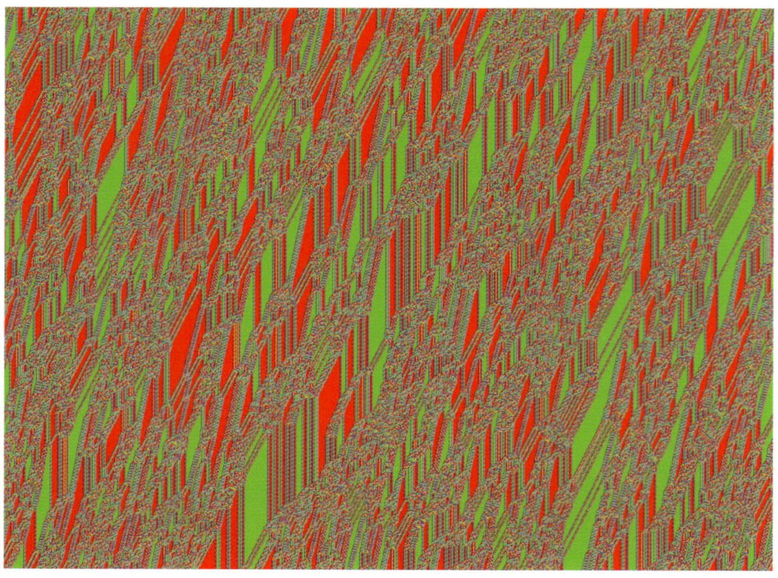

Fig. 7.8   TCOZXSHCMNMZM ($PF = 0.503, CP = 0.441$).

$2 \times 16 = 32$ possible configurations and hence $2^{32} = 4\,294\,967\,296$ rules for its next state. Patterns for this *totalistic* rule will be denoted by the code TD, two examples of which are in Figs. 7.9 and 7.10.

There are many ways to generalize these ideas to larger classes of cellular automata even in one dimension. For example, you can consider larger neighborhood sizes (larger *radius*), more than two allowed values for the states (often called *colors*), and earlier past values of the states (higher *orders*). All of these extensions greatly expand the size of the rule space. In particular, there are k^{k^m} rules for a first-order cellular automaton with m neighbors and k states. However, it is not necessary to search the entirety of the rule space, but you can just randomly sample it since many rules produce identical or similar results.

For example, suppose you need to search a rule space with 1000 bits. Rather than generate a 1000-bit number, you can construct a string of 1000 values of zeros and ones, choosing each element of the string using a random number generator (RNG). Then if you store the seed used for the random number generator, you can reconstruct the rule string at will. If the seed consists of a 32-bit number, you will be able to sample about 4×10^9 of the $2^{1000} \approx 1 \times 10^{301}$ possible rules, which is a negligible fraction, but still an immense and representative sample. The procedure is similar to how a large population of the public is randomly sampled to predict the outcome of a political election. The sample should be large enough to be representative but only needs to include a minuscule fraction of the set.

7.3   Selection Methods

Most of the patterns resulting from this enormous rule space are relatively bland and not very elegant. Thus it is useful to have ways to eliminate the uninteresting cases. Many of the methods used in the previous chapters do not work well or require modification for cellular automata.

Checking for boundedness is useless because the variables can only take values of zero and one. Checking for periodicity is similarly difficult because you need to record a long sequence of bits such as an entire row of 1200 cells, and look for the same bit pattern in a previous row but not necessarily at the same position. If the bit pattern ever repeats, it will continue to repeat forever, and the resulting image is not likely to be interesting, although it is possible for interesting structure to coexist with a periodic region if the two regions are spatially separated.

 Elegant Fractals

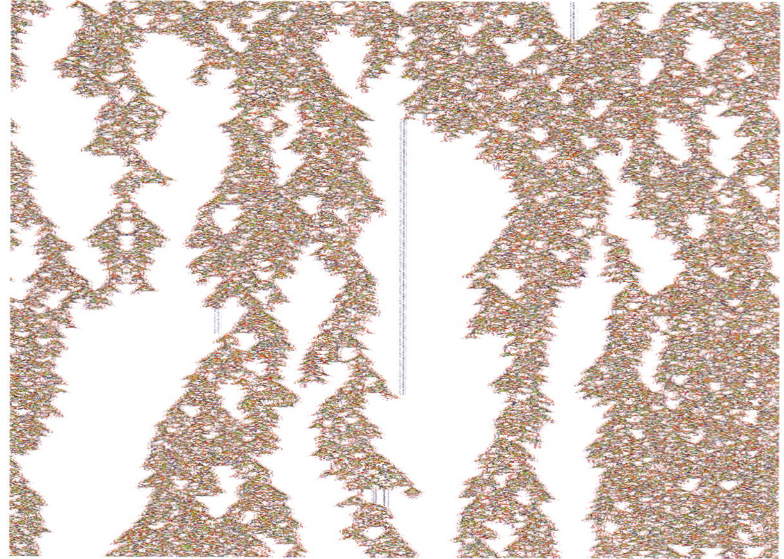

Fig. 7.9   TDEYVHONEQJDM $(PF = 0.412, CP = 0.307)$.

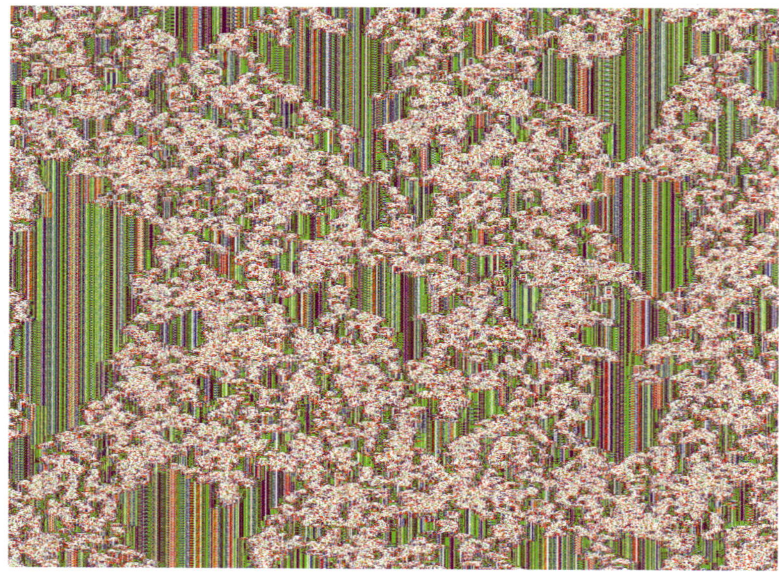

Fig. 7.10   TDLHXVTOUFWLM $(PF = 0.706, CP = 0.488)$.

Calculation of the pixel fraction (PF) is more useful if it is defined as the fraction of the pattern that are ones. If $PF = 1$ or $PF = 0$, the pattern has evolved to a spatially-uniform fixed state that can be discarded. Values close to zero or one such as $PF < 0.03$ or $PF > 0.97$ are poor candidates for the same reason. A value of PF very close to 0.5 (say $0.49 < PF < 0.51$) is also an unlikely candidate since it is probably just a simple periodic oscillation or a totally random collection of zeros and ones. Thus a value like $PF \approx 1/3$ (or $2/3$) might be a favorable indication as for the images in the previous chapters.

The cluster probability (CP), which is the probability that an arbitrary cell is the same as its four nearest neighbors (in the space-time plot), is a more useful quantity. As with the earlier attractors, intermediate values ($CP \approx 1/3$) are best. Values of PF and CP are given in the figure captions, so that you can judge for yourself how useful they are.

The skewness (SK) is not useful because it has essentially the same information as the pixel fraction since an image with PF close to zero or one will have a large value of SK, and one with PF close to 0.5 will have SK close to zero. Unfortunately, these methods eliminate only a small fraction ($\lesssim 20\%$) of the cases found, and so you will need to sort through the remaining ones yourself to find those that are elegant or invent some better criterion for automatic selection.

It is possible to estimate a Lyapunov exponent and a fractal dimension for cellular automata, but their relation to the elegance of the patterns is less clear. That would make a good study to parallel similar studies for strange attractors [Sprott (1993a)] and iterated function systems [Sprott (1994b)] if it interests you.

7.4 Two-dimensional Cellular Automata

All the methods and considerations previously described apply also to cellular automata in two dimensions. They provide a wider variety of images and better model many natural processes such as weather and landscape patterns. However, they are more computationally intensive since they require evolving a two-dimensional array ($1200 \times 900 \approx$ one million points) rather than just a single line of 1200 points. Consequently, the amount of computation required to advance the previous one-dimensional examples for 9000 generations will only advance these two-dimensional cases for nine generations. As a result, the patterns depend strongly on the initial

Moore   von Neumann

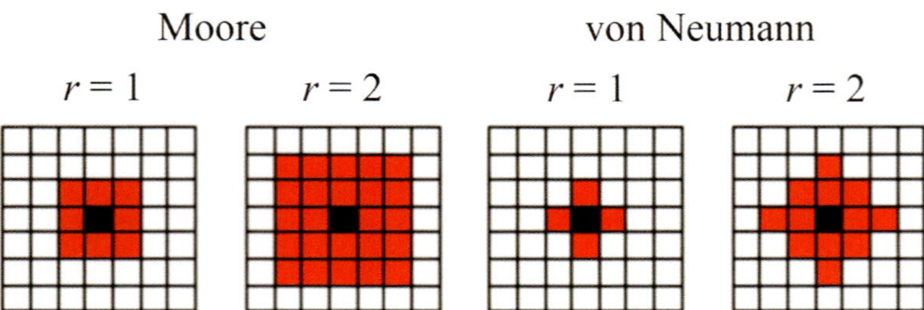

Fig. 7.11 Neighborhoods used in two-dimensional cellular automata.

conditions, and the resulting patterns do not represent attractors since they have not evolved sufficiently.

There is also the matter of how to choose a two-dimensional neighborhood. Two different methods have been described, one due to Moore (1962), and the other due to von Neumann (1966) as shown in Fig. 7.11. Clearly the number of neighbors in two dimensions can become quite large even for a small neighborhood (small r). To simplify the programming, we consider a von Neumann neighborhood with $r = 1$ since it has four neighbors and the rules used for the one-dimensional case **TA** can be employed without change.

However, it is necessary to keep two million-element arrays in memory if the sites are updated synchronously. Actually, you can do slightly better by keeping the entire current pattern and only several rows of the previous one, but the additional bookkeeping makes that less attractive. Even simpler would be to update the sites randomly, but that will not be done here since computer memory is cheap. Initial conditions and colors are chosen just as with the 1-D cellular automata.

Patterns produced by this modification of case **TA** are denoted by a code **TE**. You can also restrict the 2-D rules to symmetric and antisymmetric cases as was done in 1-D, and those will be denoted by **TF** and **TG**, respectively. Finally, you can use a totalistic rule in 2-D, which in this case is taken as a Moore neighborhood with $r = 1$, so that only an 18-bit rule is needed (nine possible values for the total value of eight neighbors times two possible values for the site at the center). A code of **TH** will be used for two-dimensional totalistic cellular automata.

Two examples of 2-D cellular automata are shown in Figs. 7.12 and 7.13. Figure 7.12 began with a random initial condition, and after only

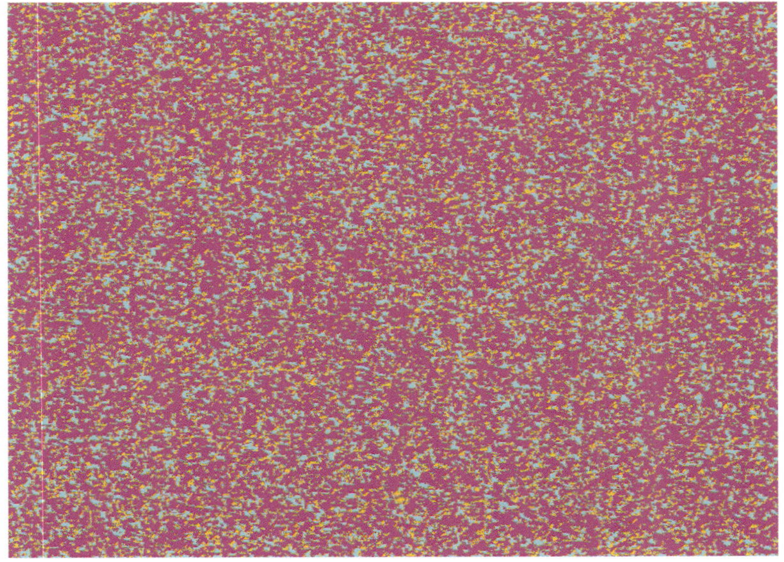

Fig. 7.12   TGLUFGCRLSHQM ($PF = 0.315, CP = 0.492$).

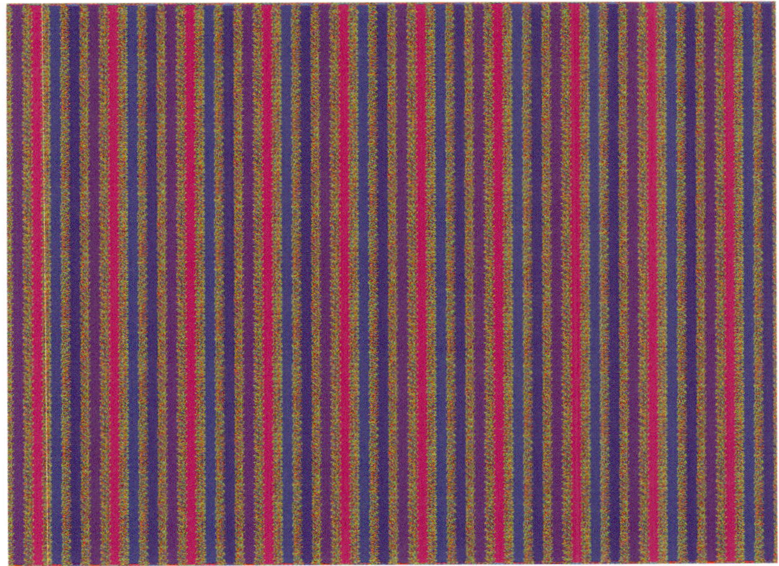

Fig. 7.13   THATNSGKUCDRM ($PF = 0.566, CP = 0.368$).

nine generations, self-organized *coherent structures* are beginning to form, reminiscent of the formation of galaxies out of the primordial gases in the early Universe. By contrast, Figure 7.13 shows a totalistic case that began with ten perfectly periodic cycles that are beginning to *disorganize* after only nine generations.

Probably the best known and most widely studied 2-D totalistic cellular automaton is the *Game of Life* developed by John Conway [Gardner (1970)]. It uses an $r = 1$ Moore neighborhood with the following rules:

(1) Any live cell with fewer than two live neighbors dies of isolation.
(2) Any live cell with two or three live neighbors lives to the next generation.
(3) Any live cell with more than three live neighbors dies of overcrowding.
(4) Any dead cell with exactly three live neighbors becomes alive by reproduction.

From a random initial condition, self-organized patterns called 'blinkers,' 'gliders,' and 'spaceships' evolve, as well as a 'glider gun' that shoots off a glider every thirty generations and thus allows self-reproducing patterns on a sufficiently large grid. The resulting patterns are not very artistically elegant, and they require a lot of computation when the grid is large, and so examples are not shown here, but they are fascinating to watch evolve in real time.

7.5 Artificial Neural Networks

We already discussed artificial neural networks as a means for automating the selection of elegant fractals and showed one such example in Eq. (2.4). It is also possible to use such networks to generate fractal patterns [Sprott (1998)]. The method to be described can be considered a generalization of a one-dimensional cellular automaton except that each cell can take on a continuum of values over some range but still with a finite number of cells (here taken as 1200) and time advancing in a number of discrete steps (here taken as 9000).

As a simple example, consider a case in which there are two neurons whose linear combination determines the value of each cell based on its previous value and the previous values of its two nearest neighbors according to

Case UA:

Parameter	Used for
a_1	Set to -1.2 (**A**)
a_2 to a_8	$x_{n+1}(i) = \sum\limits_{j=-1}^{1} a_2 \tanh(a_{j+4}x(i+j)) + a_{12}\tanh(a_{j+7}x(i+j))$
a_9	Initial conditions (see text)
a_{10}	$hue = 9.5\pi(2.2 + a_{10} + x) \mod 2\pi$
a_{11}	Replace color with white if $a_{11} \leq 0$
a_{12}	See above.

The hyperbolic tangent function has the nice feature that its values are bounded in the range of -1 to $+1$. Thus it is sometimes called a *squashing function* since its argument can be anywhere in the range of $-\infty$ to $+\infty$.

Systems of this form will be prefaced with a **UA** to indicate that it is a one-dimensional artificial neural network with a hyperbolic tangent function. The next seven symbols along with the last will denote the values of the coefficients a_2 to a_8 and a_{12} in the range $-1.2 \leq a \leq 1.3$ as in the previous cases. The index i in case **UA** indicates the spatial location of the cell, and it is evaluated modulo 1200 so that the cell at site $i = 1199$ is assumed to be adjacent to the one at $i = 0$ (periodic boundary conditions).

The other coefficients are used in the same way as they were for the cellular automata except that the initial conditions can take on a continuum of values chosen according to $\cos(\pi a_9 i/60)$ for periodic initial conditions and from a uniform random value in the range of -1 to $+1$ for the random initial conditions. Also, instead of choosing the color from 32 discrete values determined by the previous neighborhood configuration, it is chosen from a continuum of hues according to the value of x in each cell using Eq. (6.1) with $h = 0.5\pi(2.2 + a_{10} + x) \mod 2\pi$.

Most of the patterns produced by this simple artificial neural network with only three inputs and two neurons for each cell are not very interesting, but a few such as the one in Fig. 7.14 are marginally elegant. Periodicity checking is useful for the 80% of the cases in which the neurons become saturated at a value of ± 1 or where they alternate periodically. Most of the remaining cases have PF close to 0.5 and small values of CP and SK.

Slightly more interesting patterns are produced if the hyperbolic tangent is replaced by the sine function,

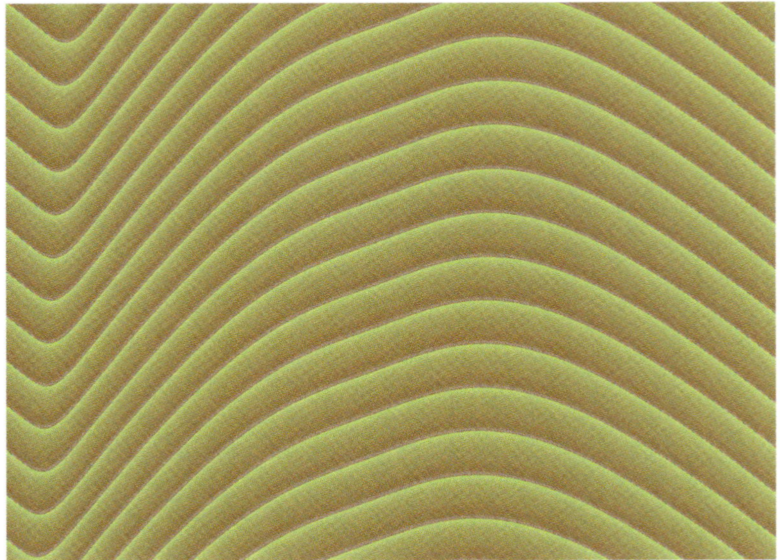

Fig. 7.14   UAEIHTBOOGXQE ($PF = 0.500, CP = 0.005$).

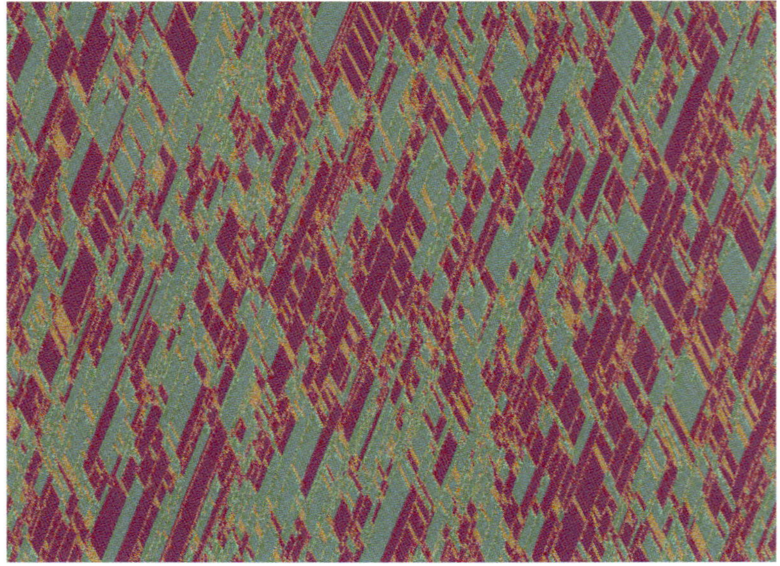

Fig. 7.15   UBDKXDBDDZQWC ($PF = 0.526, CP = 0.514$).

Case UB:

Parameter	Used for
a_1	Set to -1.1 (B)
a_2 to a_8	$x_{n+1}(i) = \sum_{j=-1}^{1} a_2 \sin(a_{j+4}x(i+j)) + a_{12}\sin(a_{j+7}x(i+j))$
a_9	Initial conditions (see text)
a_{10}	$hue = 9.5\pi(2.2 + a_{10} + x) \mod 2\pi$
a_{11}	Replace color with white if $a_{11} \leq 0$
a_{12}	See above,

which is also bounded in the range -1 to $+1$ but has positive and negative derivatives and does not saturate as does the hyperbolic tangent. An example produced in this way is in Fig. 7.15.

Other functions are also possible. While the sine function is odd (antisymmetric about zero), the cosine function is even (symmetric about zero) giving a system of the form

Case UC:

Parameter	Used for
a_1	Set to -1.0 (C)
a_2 to a_8	$x_{n+1}(i) = \sum_{j=-1}^{1} a_2 \cos(a_{j+4}x(i+j)) + a_{12}\cos(a_{j+7}x(i+j))$
a_9	Initial conditions (see text)
a_{10}	$hue = 9.5\pi(2.2 + a_{10} + x) \mod 2\pi$
a_{11}	Replace color with white if $a_{11} \leq 0$
a_{12}	See above,

with an example solution as shown in Fig. 7.16.

Furthermore, you can have a combination of sine and cosine,

Case UD:

Parameter	Used for
a_1	Set to -0.9 (D)
a_2 to a_8	$x_{n+1}(i) = \sum_{j=-1}^{1} a_2 \sin(a_{j+4}x(i+j)) + a_{12}\cos(a_{j+7}x(i+j))$
a_9	Initial conditions (see text)
a_{10}	$hue = 9.5\pi(2.2 + a_{10} + x) \mod 2\pi$
a_{11}	Replace color with white if $a_{11} \leq 0$
a_{12}	See above,

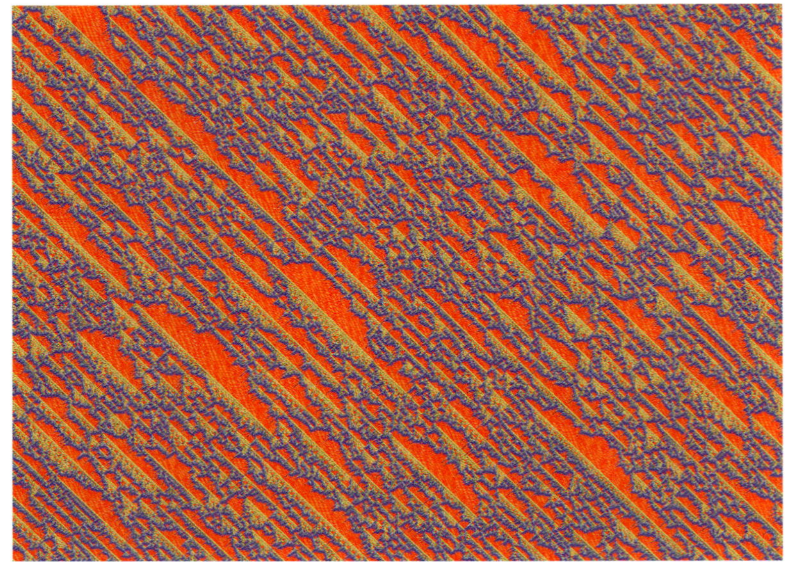

Fig. 7.16 **UCXVQABDJRPXE** $(PF = 0.710, CP = 0.617)$.

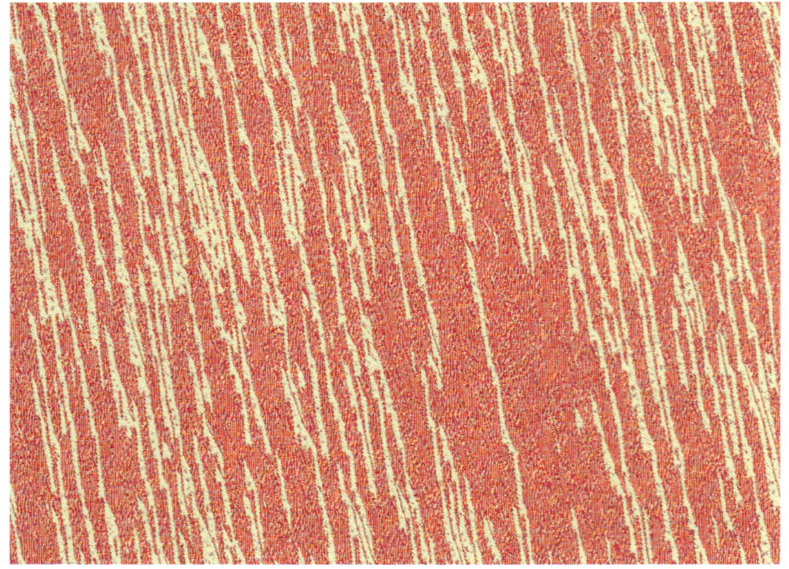

Fig. 7.17 **UDXEBQRHHCVLY** $(PF = 0.706, CP = 0.412)$.

with an example solution as shown in Fig. 7.17. More elegant versions of these systems should be possible by using more than two neurons, a neighborhood larger than the nearest two, and especially time delays greater than one (higher orders).

7.6 Coupled Map Lattices

The artificial neural networks just described are part of a larger class of system called a *coupled map lattice* [Kaneko (1984); Waller and Kapral (1984); Kuznetsov and Pikovsky (1985)]. Just like the iterated maps in Chapter 3, they advance a continuous variable in discrete time steps, but they do so with many such variables simultaneously, each coupled to its neighbors. We will consider a simple case with six neighbors given by

Case UE to UM:

Parameter	Used for
a_1	Set to -0.8 to 0 (E to M)
a_2 to a_8	$x_{n+1}(i) = \sum\limits_{j=-3}^{3} f(a_{j+5} x(i+j))$
a_9	Initial conditions (see text)
a_{10}	$hue = 9.5\pi(2.2 + a_{10} + x) \mod 2\pi$
a_{11}	Replace color with white if $a_{11} \leq 0$
a_{12}	See above,

where the function $f(x)$ can take on a number of different forms:

$$
\begin{aligned}
&\text{Case UE:} \quad && f(x) = \sin x \\
&\text{Case UF:} \quad && f(x) = \cos x \\
&\text{Case UG:} \quad && f(x) = 1 - 2x^2 \\
&\text{Case UH:} \quad && f(x) = 1 - 2x^4 \\
&\text{Case UI:} \quad && f(x) = 1 - 2|x| \\
&\text{Case UJ:} \quad && f(x) = 1 - (2|x+1| \mod 2) \\
&\text{Case UK:} \quad && f(x) = 1 - (4|x+1| \mod 2) \\
&\text{Case UL:} \quad && f(x) = 1 - (6|x+1| \mod 2) \\
&\text{Case UM:} \quad && f(x) = 1 - (8|x+1| \mod 2).
\end{aligned}
$$

Most of the cases are automatically bounded in the range of -1 to $+1$, so that they map the interval $(-1, 1)$ back onto itself (an *endomorphism*),

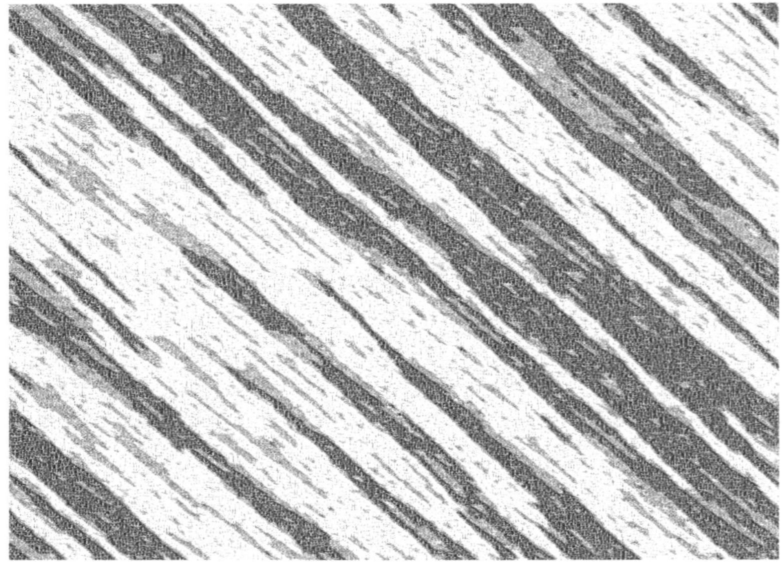

Fig. 7.18 **UFZTPEPUTLEAM** $(PF = 0.520, CP = 0.537)$.

which typically allows chaotic solutions. However, cases **UG**, **UH**, and **UI** are not always bounded. These simple systems produce many chaotic solutions, but very few show the kind of intricate spatial structure that makes for elegant fractals. The selection methods used for the artificial neural networks are equally (in)effective here, except that checking for boundedness is especially important. Two of the more interesting examples are shown in Figs. 7.18 and 7.19.

7.7 Coupled Flow Lattices

Just as Chapter 4 is an extension of the iterated maps in Chapter 3 to cases where time advances continuously rather than in discrete steps, we can extend the coupled map lattices to lattices of coupled ordinary differential equations. Each of the thirteen previous cases **UA** to **UM** has a corresponding *coupled flow lattice*, the first twelve of which we will label **UN** to **UY**. The case **UZ** will be reserved for a partial differential equation to be described shortly.

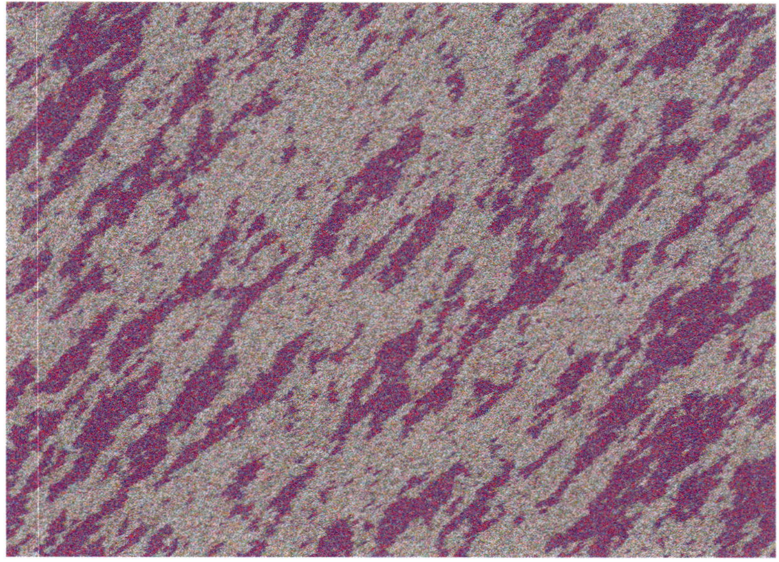

Fig. 7.19   `ULIJLVIFIBABM` ($PF = 0.848, CP = 0.646$).

All that is necessary is to calculate the value of $x_{n+1}(i)$ using the appropriate equation, but rather than use it directly to update each site, you multiply it by a small time step such as $h = 0.001$ and add it to the previous value for each site:

$$\text{Cases UN to UZ:}$$
$$x_{n+1}(i) \rightarrow hx_{n+1}(i) + x_n(i). \tag{7.1}$$

There are several problems with producing patterns in this way. Even though the calculated value of x may be bounded, when it instead plays the role of $\dot{x}$ (the time derivative of x), the resulting orbit is not necessarily bounded. Thus it is important to check for boundedness as well as periodicity. In fact, about 42% of the cases are periodic, and about 56% are unbounded, leaving only about 2% as candidates for visually interesting images, only a small fraction of which are elegant.

Even when the orbit is bounded and aperiodic, it may have very small or very large values, making it difficult to choose appropriate colors. Furthermore, because h is a small number, the orbit advances very slowly, so that 9000 iterations only moves it forward nine time units for $h = 0.001$. Thus the result may reflect the initial conditions more than the final attracting state. Consequently, many of the patterns, which show only the time interval of 8.1 to 9.0, are rather bland with a nearly uniform color or with simple vertical stripes. However, occasional interesting patterns emerge such as those in Figs. 7.20 and 7.21.

7.8   Partial Differential Equations

The most general spatiotemporal models are described by *partial differential equations* (PDEs) in which space, time, and the value of the dependent variable(s) all have a continuum of possible values. As an example, one of the simplest such model [Brummitt and Sprott (2009)] that exhibits spatiotemporal chaos even in one spatial dimension is the *Kuramoto–Sivashinsky equation* [Kuramoto and Tsuzuki (1976); Sivashinsky and Michelson (1980)],

$$\frac{\partial x}{\partial t} = -x \frac{\partial x}{\partial s} - \frac{\partial^2 x}{\partial s^2} - \frac{\partial^4 x}{\partial s^4}, \tag{7.2}$$

where s is the spatial position displayed horizontally while time increases downward in accord with the previous cases. As with the other coupled flow lattices, the boundary conditions are taken as periodic (the right edge of the image is assumed to be adjacent to the left edge).

The second-derivative term $\frac{\partial^2 x}{\partial s^2}$ is a negative *viscosity* leading to the growth of long-wavelength modes, and the fourth-derivative term $\frac{\partial^4 x}{\partial s^4}$ is a hyperviscosity that damps the short-wavelength modes. The nonlinear term $u \frac{\partial x}{\partial s}$ transports energy from the growing modes to the damped modes.

Of course, a digital computer always must discretize the variables, and the usual way to do that is to convert the partial differential equation into a coupled flow lattice in which the spatial derivatives are represented by coupling to adjacent lattice sites,

$$\frac{\partial x}{\partial s} = (x_{i-2} - 8x_{i-1} + 8x_{i+1} - x_{i+2})/12k$$

$$\frac{\partial^2 x}{\partial s^2} = (-x_{i-2} + 16x_{i-1} - 30x_i + 16x_{i+1} - x_{i+2})/12k^2 \tag{7.3}$$

$$\frac{\partial^4 x}{\partial s^4} = (x_{i-2} - 4x_{i-1} + 6x_i - 4x_{i+1} + x_{i+2})/k^4,$$

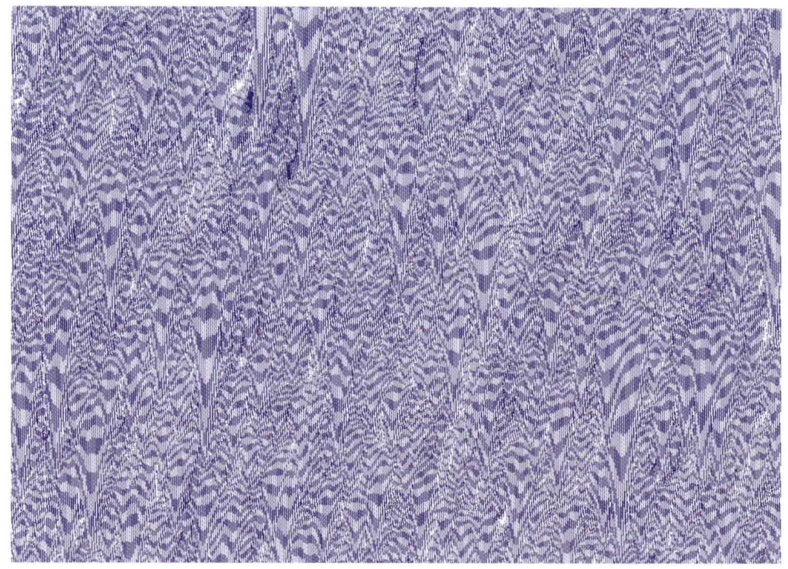

Fig. 7.20 **UXOELDNSUFUMM** $(PF = 0.500, CP = 0.018)$.

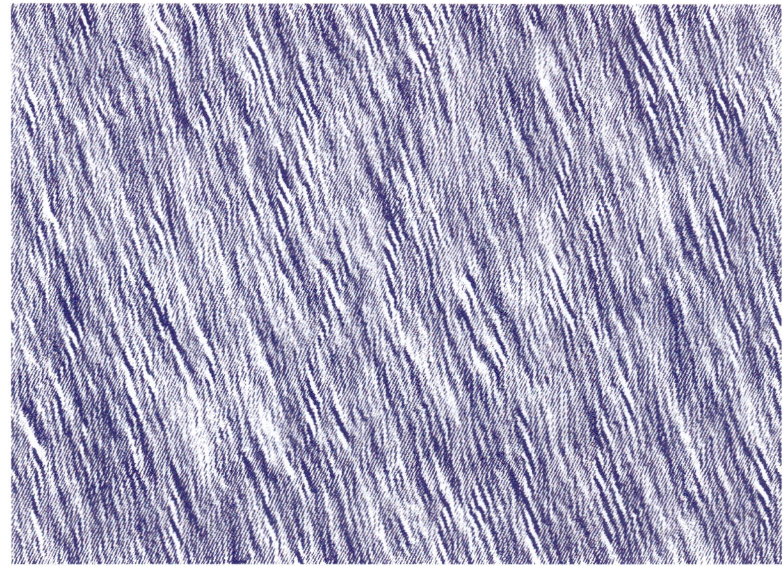

Fig. 7.21 **UYVJMAXQCOPEM** $(PF = 0.512, CP = 0.324)$.

where k is a small constant that plays the same role for the spatial derivatives as h did for the time derivatives in the ordinary differential equations.

Equation (7.2) can be solved as a special case of the more general coupled flow lattice given by

Case UZ:

Parameter	Used for
a_1	Set to 1.3 (Z)
a_2 to a_{11}	$x_{n+1}(i) = x(i) + \frac{h}{k} \sum_{j=-2}^{2} (a_{j+4}x(i) + a_{j+9})x(i+j)$
a_{10}	$hue = 9.5\pi(2.2 + a_{10} + x) \mod 2\pi$
a_{11}	Replace color with white if $a_{11} \le 0$
a_{12}	Initial conditions (see text),

where h and k are both small constants whose ratio is taken as $h/k = 0.1$, which gives a similar number of periodic and unbounded solutions, neither of which provide elegant images. Usually chaos occurs when the parameters are such that a system is just on the verge of instability. Any of the cases that pass the periodicity and boundedness tests are worth examining, but they constitute only about one of every thousand cases tested. The pixel fraction and cluster probability are not very useful since they do not vary much for the cases found. The parameter a_{12} is used to specify the initial conditions in the same way that a_9 was used for the previous coupled map and coupled flow lattices.

Of course, this is just one of an unlimited variety of possible nonlinear partial differential equations, but it is one with a different type of nonlinearity than in the previous cases, involving the product of the values of the variable at different spatial positions. It is known to have chaotic solutions, and thus it is a good candidate for producing elegant fractal patterns, five examples of which are in Figs. 7.22 to 7.26, prefaced with UZ to indicate that they came from Eq. (7.3).

All of these one-dimensional cases have two-dimensional and higher-dimensional counterparts as with the cellular automata, but their calculation requires considerable computer time and typically does not produce images any more elegant than the one-dimensional cases shown here, and so that will be left for you to explore.

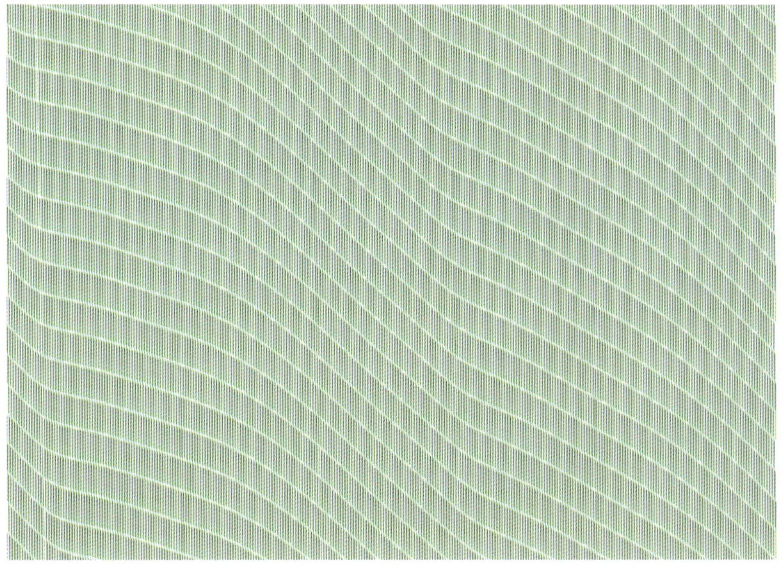

Fig. 7.22 **UZZPPMUDQTHFB** $(PF = 0.478, CP = 0.000)$.

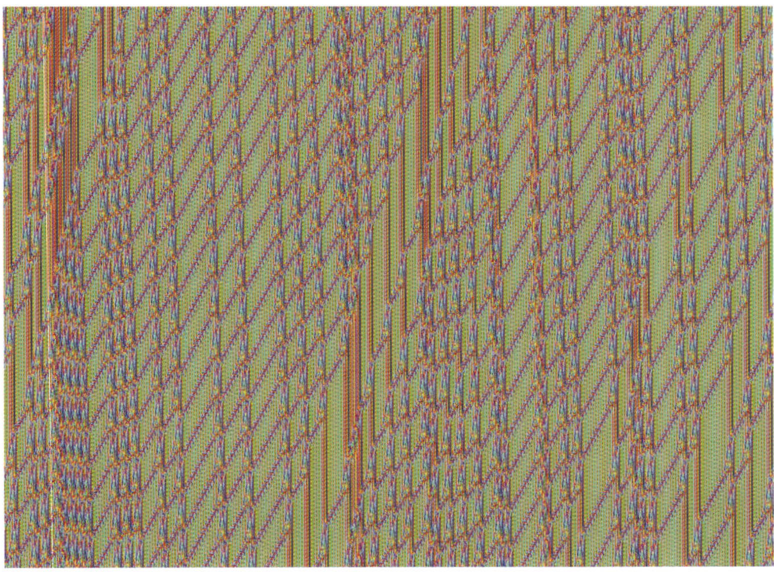

Fig. 7.23 **UZJIKGIRAOGOO** $(PF = 0.488, CP = 0.000)$.

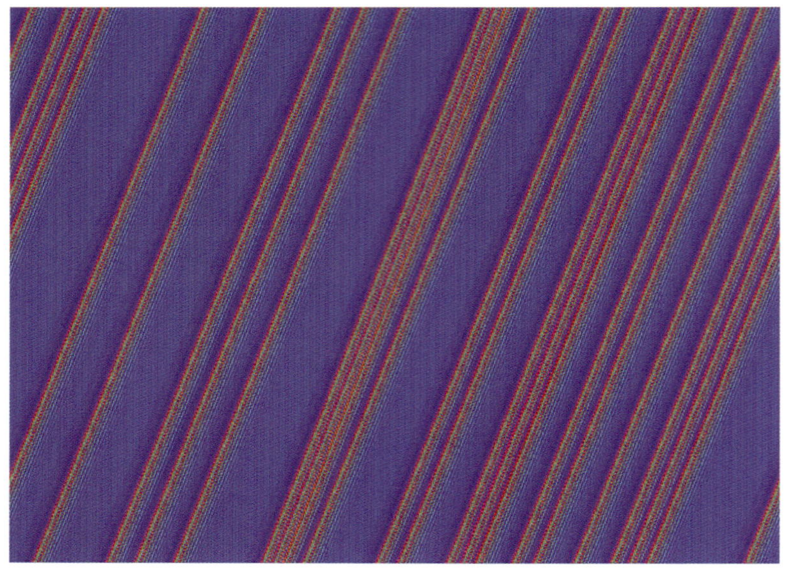

Fig. 7.24   UZSVPVGIFFIWN $(PF = 0.497, CP = 0.000)$.

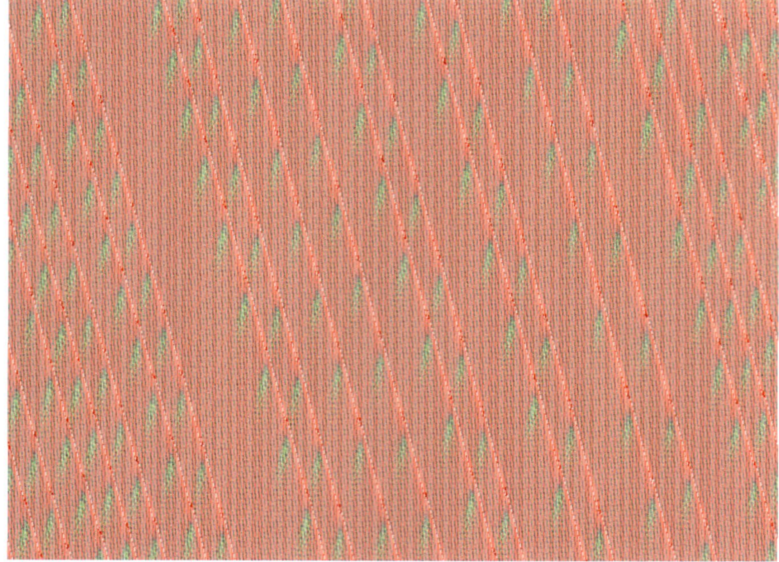

Fig. 7.25   UZMTMPUSELOJZ $(PF = 0.498, CP = 0.002)$.

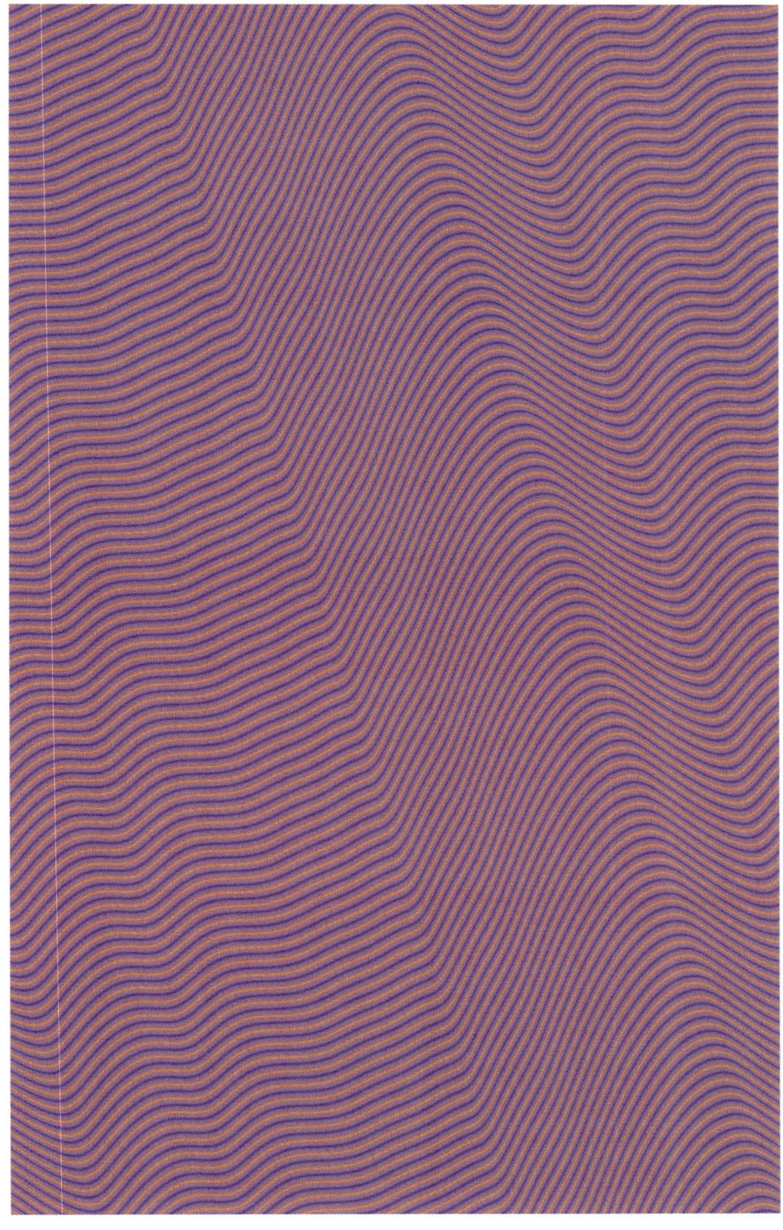

Fig. 7.26 UZXMNOWCOMMRV $(PF = 0.410, CP = 0.000)$.

Chapter 8

Random Fractals

All the previous examples are from deterministic dynamical systems, although the rules, initial conditions, and colors are sometimes chosen randomly. As a result, the same initial conditions will give exactly the same pattern every time, and the resulting fractal patterns are perfectly self-similar. Here we consider stochastic cases where the rules involve randomness and the resulting fractals are only statistically self-similar, and sometimes not even that.

8.1 Random Number Generators

To produce random fractals, it is necessary to have a source of random numbers. That raises the question of how a computer performing simple arithmetic operations can be a source of randomness. Of course, it cannot, but it can produce a sequence of numbers that pass all tests for randomness and that lack any discernible pattern even though we know that the numbers came from a deterministic algorithm. Numbers produced in this way are called *pseudorandom* to distinguish them from true randomness.

Most computer languages have a built-in pseudorandom number generator (RNG), and the method used to produce the numbers is usually undocumented and sometimes proprietary and secret. Typically, it will produce a sequence of floating-point numbers with a fixed precision in the range of 0 to 1, and the sequence can be replicated exactly by specifying the *seed* for the random number generator. Often the seed is chosen from the time of day when the program is launched so that a different sequence results unless the program happens to be launched at exactly the same second on different days. Alternately, if you want a seed that is unique to you, use your telephone number or (American) Social Security number.

Generally, programs written in different computer languages will produce a different random sequence even with the same seed since they use different algorithms. Many of these generators, especially in the early days of computers, had serious flaws, some obvious, others less so. For our purposes, it is useful to have a reasonably good random number generator and one that is *portable*, which is to say one that can be implemented in any computer language and that will produce the same sequence of numbers from a given seed in each case.

The simplest and usual type of random number generator is the *multiplicative linear congruential generator* (LCG), which is nothing more than a one-dimensional iterated map of the form

$$r_{n+1} = ar_n \mod m, \tag{8.1}$$

where r, a, and m are all integers and m is a prime number. This produces a reproducible pseudorandom sequence of integers r_n for any given seed r_0. Then to get a pseudorandom number ran in the range of 0 to 1, you simply divide the corresponding integer iterate by m to obtain $ran = r/m$.

The parameters a and m should be chosen such that every integer r in the range of 1 to $m - 1$ occurs exactly once, after which the sequence will repeat. This guarantees that the random numbers are uniform over the range of $0 < ran < 1$ after dividing the integers by m. Recommended values [Park and Miller (1988)] are $a = 7^5 = 16\,807$ and $m = 2^{31} - 1 = 2\,147\,483\,647$, which will produce over two billion different values. That sounds like a large number, but with computers that run in the GHz (10^9 cycles per second) range, the numbers could potentially be exhausted in a matter of seconds.

Note that Eq. (8.1) is nothing more than a chaotic mapping with a large Lyapunov exponent of $\log a \approx 9.73$, which effectively guarantees that the subsequent iterates of two nearly equal seeds have values that are essentially uncorrelated. You can think of the multiplication as moving the bits of the number r to the left by an amount equal to the integer part of $\log_2 a \approx 14$, discarding the bits that are replaced, and bringing in new bits on the right that represent the fractional part of the multiplication. The lost bits represent the loss of information about the initial condition that is the hallmark of chaos.

In addition to the short cycle time, successive values are not completely independent of one another. For example, if one of the random numbers has a small value, say less than 1×10^{-6}, which should occur on average every million iterations, the next value will always be less than $10^{-6}a$, or 0.016807 for the above values.

More generally, if k successive iterates are plotted in a k-dimensional space, they will lie on $(k-1)$-dimensional hyperplanes, and there will be at most about $\sqrt[k]{m}$ such planes. Thus for triplets of points ($k = 3$), there will be at most $\sqrt[3]{2\,147\,483\,647} \approx 1290$ planes, which may be far too few for many purposes.

Furthermore, note that you cannot use a seed of $r_0 = 0$, which might seem to be a natural choice, since that will simply lead to an infinite sequence of zeros. That problem can be circumvented by adding a constant c,

$$r_{n+1} = (ar_n + c) \mod m, \tag{8.2}$$

with recommended values [Knuth (1998)] of $a = 7141$, $c = 54773$, and $m = 259\,200$, which does not overflow with 32-bit arithmetic but with a short periodicity of $m = 259\,200$.

Many of these problems can be circumvented by *shuffling* [Knuth (1997)]. Suppose you make a table of 100 imperfect random numbers. Then each time you need a new number, you use the previous number to decide which entry in the table to use, and then you replace it with a new imperfect random number. This method can be applied to any random number generator including the one built into whatever programming language you are using, and it should produce sequences of numbers that are sufficiently random for nearly all purposes.

However, we desire a *portable* random number generator that will produce the same sequence from a given seed no matter what programming language or type of machine you are using. For that purpose, a method proposed by L'Ecuyer (1998) is useful. The idea is to generate two random integers using different values of the parameters, and then to obtain the final number by subtracting one from the other and adding a fixed constant if the result is negative.

Suggested values of m for the two generators are $m_1 = 2\,147\,483\,563$ and $m_2 = 2\,147\,483\,399$. Since $m_1 - 1$ and $m_2 - 1$ share only the factor 2, the period of the combined generator is $m_1 m_2 / 2 \approx 2.3 \times 10^{18}$. When combined with shuffling, the resulting numbers are so random that Press and Teukolsky (1992) once offered \$1000 to the first person who could convince them otherwise. The prize was never awarded, and the offer was eventually withdrawn because most claimants did not understand what constitutes a statistical proof of randomness [Press *et al.* (2007)]. For more details about the method, see the program *ran2* in the second edition of *Numerical Recipes*.

8.2 Random Walks

Armed with a good random number generator, we can explore the simplest example of a random fractal, which is just a *random walk*. Starting from some initial condition in a six-dimensional space, at each iteration we use six random numbers $ran_1, ran_2, \ldots, ran_6$ and move the orbit a small distance in the range of $-h$ to $+h$ with $h = 0.001$ in each of the six dimensions to a new position given by

$$\text{Case V:}$$
$$
\begin{aligned}
x_{n+1} &= x + h(2ran_1 - 1) \\
y_{n+1} &= y + h(2ran_2 - 1) \\
z_{n+1} &= z + h(2ran_3 - 1) \\
u_{n+1} &= u + h(2ran_4 - 1) \\
v_{n+1} &= v + h(2ran_5 - 1) \\
w_{n+1} &= w + h(2ran_6 - 1).
\end{aligned}
\tag{8.3}
$$

Other forms for the quantity in parenthesis are often taken such as Gaussian random numbers with zero mean and unit variance, or simply ± 1 with the sign chosen randomly, which can be calculated from $\mathrm{sgn}(ran - 0.5)$. The resulting patterns are similar, and so the simple form in Eq. (8.3) will be used.

As with the other iterated maps in Chapter 3, we project the orbit onto the xy-plane with depth clues determined by z and with color intensities given by $red = 128(u + 1), green = 128(v + 1)$, and $blue = 128(w + 1)$. Since there is no way to know in advance how far the orbit will wander from its initial position, we take a region bounded by ± 1 in each dimension and apply periodic boundary conditions (if a variable is less than -1, add 2; if it is greater than $+1$, subtract 2). Initial conditions for all variables are taken as 0.1. The color of the orbit begins as a light gray since u, v, and w all start small and equal, but they quickly walk away toward a more pleasing color which changes abruptly whenever one of the color variables (u, v, w) exceeds 1.0 or falls below zero.

The mean square displacement in the plane at each time step is given by $\langle x^2 \rangle + \langle y^2 \rangle = 2h^2/3$, and thus we expect the orbit to wander after N iterations an average distance of $h\sqrt{2N/3} \approx 2.3$ for $N = 1 \times 10^7$. Thus the orbit should usually traverse the region of width 2.0 at least once and sometimes several times.

Figures 8.1 and 8.2 show two examples obtained in this way with a code prefaced with a V to indicate that it is a random walk from Eq. (8.3). The second letter of the code (A) indicates that the colors were determined as described above, and the next six letters of the code are used to seed the

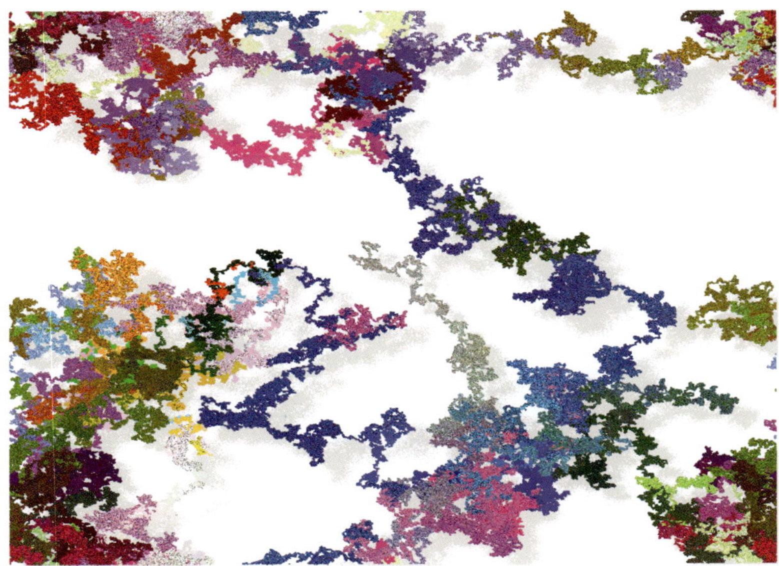

Fig. 8.1 **VAWTZMPBMMMMM** $(PF = 0.387, CP = 0.782, SK = -0.041)$.

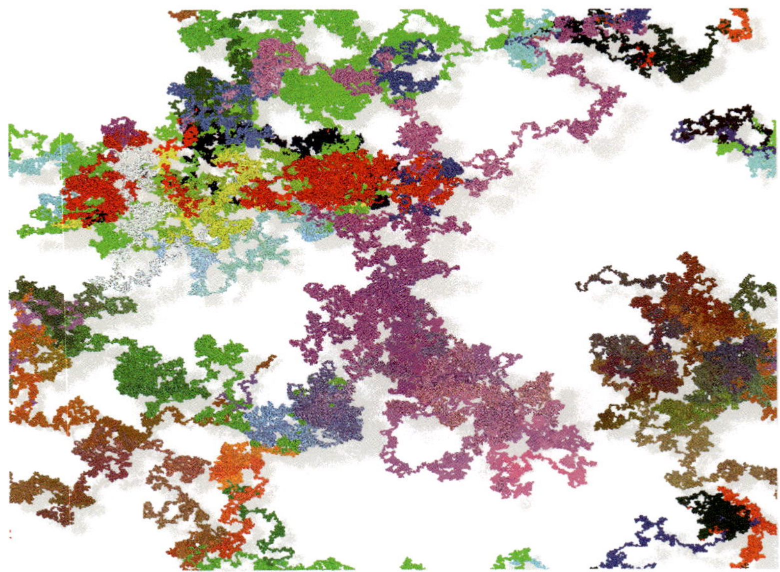

Fig. 8.2 **VAQYHUOMMMMMM** $(PF = 0.387, CP = 0.779, SK = -0.116)$.

random number generator so that the image can be replicated identically. The remaining five letters are not used and are set to M (zero).

All the random walks have a certain sameness about them. They mainly differ in how uniformly they fill the bounding rectangle and how the colors happen to come out. In particular, they tend to have similar values for the pixel fractal ($PF \approx 1/3$) and cluster probability ($CP \approx 0.8$). The pixel fractal is a measure of how much the random walk tends to avoid over-writing itself, and h and N were chosen so that PF has a value close to the optimum of $PF = 1/3$ for other image types. The cluster probability has a small variance and contains much the same information as the pixel fraction, although you could eliminate the largest values of CP since they all exceed the value of $CP = 1/3$ that was previously shown to be optimal. Consequently, these quantities are not very useful in selecting elegant examples.

Somewhat more useful is the skewness, which is a measure of the horizontal symmetry. For example, you could eliminate approximately half of the cases, those with $|SK| > 0.5$ and be fairly sure that the remaining half are worth examining. Even easier would be just to look at the average value of the horizontal coordinate $\langle x \rangle$ and eliminate any cases for which it exceeds some small value such as $|\langle x \rangle| > 0.1$. As mentioned earlier, horizontal asymmetry is more aesthetically objectionable than vertical asymmetry, although you could certainly also consider the value of $\langle y \rangle$ if you think vertical symmetry is important. You could also simply rotate the image by 90 degrees and see if it looks better to you. In the end, all the random walks look rather similar.

8.3 Random Walk Variants

As with many of the previous systems, the lack of symmetry degrades the elegance of these images. One way to restore some symmetry is to make them into symmetric icons as was done with some of the earlier cases. Figures 8.3 and 8.4 show two such examples denoted by the code VB and a digit at the end to indicate the number of petals in the icon. The previous automatic selection criteria are of limited use here because the skewness is largely hidden in the icon transformation, but nearly all the cases are worth a look.

There are also other ways to choose the colors. For example, you could easily implement the 'coloring book scheme,' producing images such as Figs. 8.5 and 8.6 denoted by the code VC and symmetric icons such as Figs. 8.7 and 8.8 denoted by the code VD.

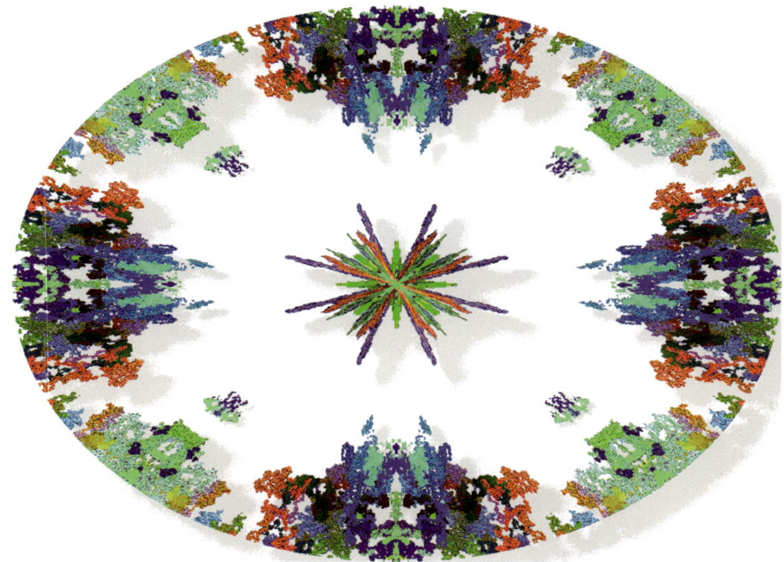

Fig. 8.3   **VBCYPZLOMMMMM8** $(PF = 0.332, CP = 0.757, SK = 0.240)$.

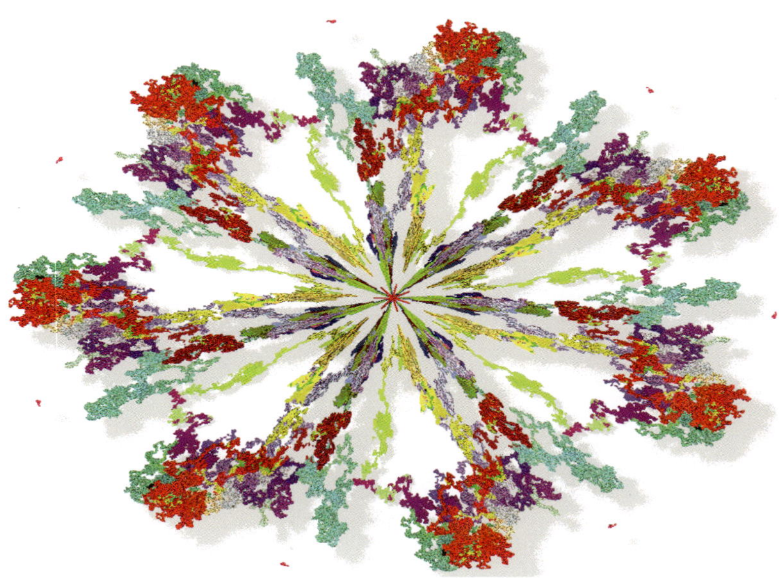

Fig. 8.4   **VBOEFVFNMMMMM7** $(PF = 0.339, CP = 0.721, SK = -0.137)$.

 Elegant Fractals

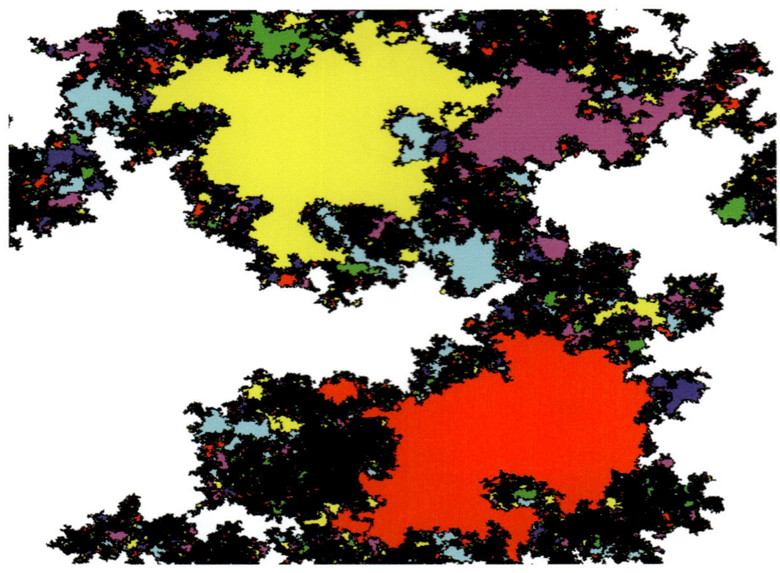

Fig. 8.5   **VCPUVMZHMMMMM** $(PF = 0.335, CP = 0.792, SK = -0.766)$.

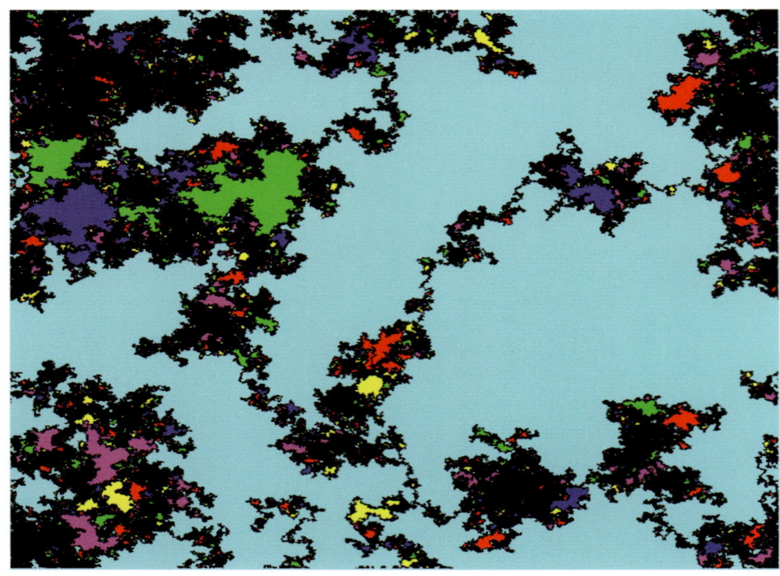

Fig. 8.6   **VCPDRCVMMMMMM** $(PF = 0.340, CP = 0.775, SK = -0.761)$.

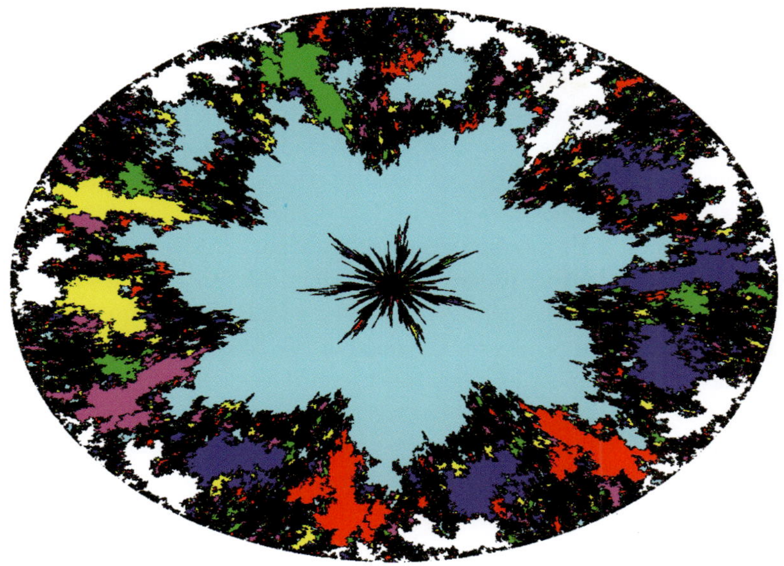

Fig. 8.7   `VDXLHXQMMMMMM7` ($PF = 0.340, CP = 0.736, SK = 0.718$).

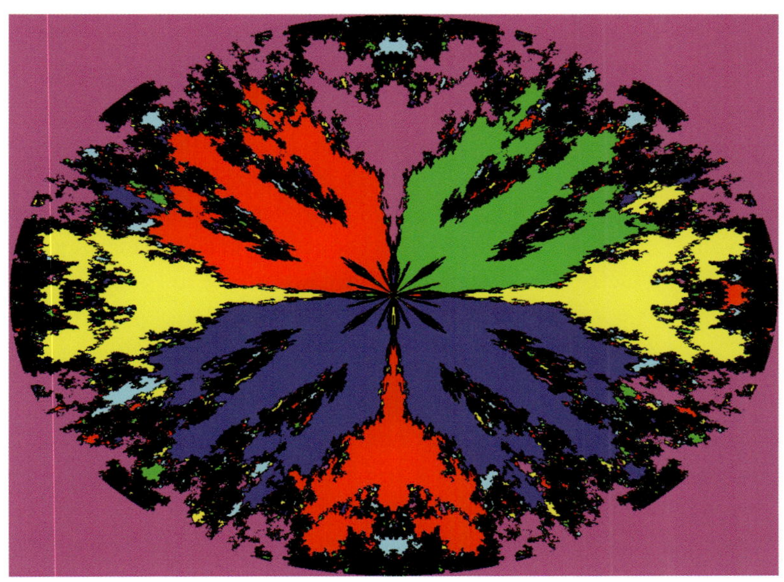

Fig. 8.8   `VBOEFVFNMMMMM7` ($PF = 0.338, CP = 0.746, SK = 1.471$).

8.4   Diffusion Limited Aggregation

A different type of random fractal is produced by a process called *diffusion limited aggregation* (DLA) [Witten and Sander (1983)]. Imagine that the air around you is filled with tiny dust particles moving randomly (*Brownian motion*). Assume that one of these particles is sticky so that another particle sticks to it when they collide, and it then also becomes sticky. The process continues until a large cluster of such particles is formed.

On the computer, this process can be simulated by starting with a single activated pixel at the center of the screen. Particles are sequentially released from random points on the edge of the screen and allowed to random walk (without plotting their path) until they touch the one at the center, whereupon they join the cluster and also become sticky. If the particle walks off the edge of the screen, it instantly reappears at the opposite edge just as for the random walks previously described. Two particles are assumed to collide when they are in the $r = 1$ von Neumann neighborhood of one another (see Fig. 7.11), although other choices are possible. The cluster is not allowed to move, or perhaps you can imagine the calculation taking place in a frame of reference moving with the cluster.

Since the process is very slow, the random walk is accelerated by having the particles move randomly one pixel up, down, right, or left with each iteration, and the process is stopped when the cluster grows to touch one of the boundaries. Even so, the calculation is an order of magnitude slower than most of the previous cases.

The customary way to color the cluster is according to the sequence with which the points were attached to it, for example in a hue that is proportional to that number N using Eq. (6.1) to determine the corresponding (r, g, b) values. To give more variety to the colors, the starting hue and the proportionality factor are chosen randomly according to

Case W:

Parameter	Used for
a_1	Type of plot (see text) (A, B, C, D)
a_2 to a_8	Seed for random number generator
a_9 and a_{10}	$hue = 0.8\pi(1.2 + a_{10} + 10^{-4}a_9 N)$
a_{11}	Sticking probability $= (1.3 + a_{11})/2$
a_{12}	Initial and boundary conditions, 0, ∓ 1 (M, C, W).

In addition, the particles are assumed to stick with a probability between 0.05 and 1 given by $P = (1.3 + a_{11})/2$. These cases are indicated by a code beginning with WA, two examples of which are in Figs. 8.9 and 8.10. Figure

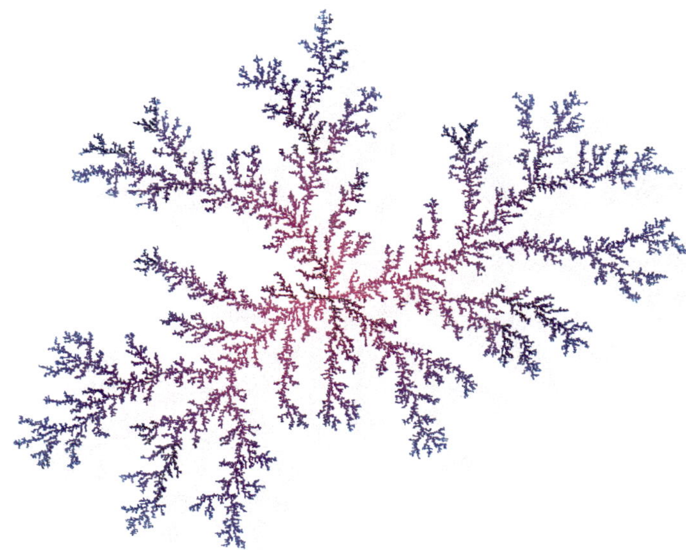

Fig. 8.9 **WAKAPNTGRNZPM** $(PF = 0.100, CP = 0.045, SK = 0.000)$.

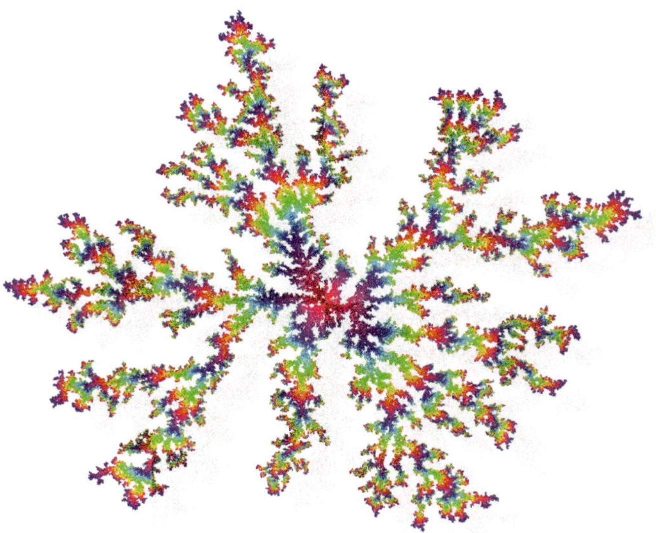

Fig. 8.10 **WAEOQIIMURPAM** $(PF = 0.159, CP = 0.317, SK = 0.000)$.

8.9 has a high sticking probability of 0.8, and Fig. 8.10 has a low sticking probability of 0.05. Cases with a low sticking probability take longer to calculate and produce a less 'delicate' image. DLA patterns are also known as *Brownian trees* or *Lichtenberg figures*, and they are often associated with electrical discharges such a lightning bolts.

Since these cases calculate rather slowly, you will not have the luxury to make a large collection of them, but that is not a serious limitation since they are all somewhat similar. The main differences are in the colors and the thickness of the plot as determined by the sticking probability. The pixel fraction and cluster probability are similar for cases with the same sticking probability. The skewness is useful to eliminate the small number of highly asymmetric cases, although most cases have negligible skewness.

8.5   DLA Variants

As with the random walks, the inherent asymmetry of the DLA clusters detracts from their elegance. Some elegance can be restored by plotting them as symmetric icons, two examples of which are in Figs. 8.11 and 8.12, denoted by a second letter of B rather than A in their code. Figure 8.11 has a sticking probability of 0.65, and Fig. 8.12 has a sticking probability of 1.0. In this case, the individual petals of the icon do not need to be highly symmetric, and the examples shown have significant skewness.

You can also use the 'coloring book scheme' to fill in the enclosed white spaces, which are relatively rare in DLA clusters, especially those with a high sticking probability. The main effect is to plot the cluster in black and to add a background color. Two such examples are shown in Fig. 8.13 with a sticking probability of 0.05 and Fig. 8.14 with a sticking probability of 0.15. A second letter in the code of C denotes a regular plot, and a D denotes a symmetric icon, both with the alternate coloring scheme.

You can grow the cluster from objects other than a single point at the center of the screen such as a collection of points distributed randomly or in some pattern or on a circle or segment of a line or multiple lines or perhaps some alphabetic characters. Figure 8.15 shows a case where the rectangular boundary surrounding the image is sticky, and particles are started from the center of the rectangle. This variant will be denoted by a value of $a_{12} = -1$ (C at the end of the code) rather than $a_{12} = 0$ (M) as in the previous DLA examples. Figure 8.16 shows a case where the sticky boundary is a circle denoted by a value of $a_{12} = +1$ (W at the end of the code). These systems can also be displayed as icons, examples of which are in Figs. 8.17 and 8.18 with a trailing digit at the end of the code to indicate

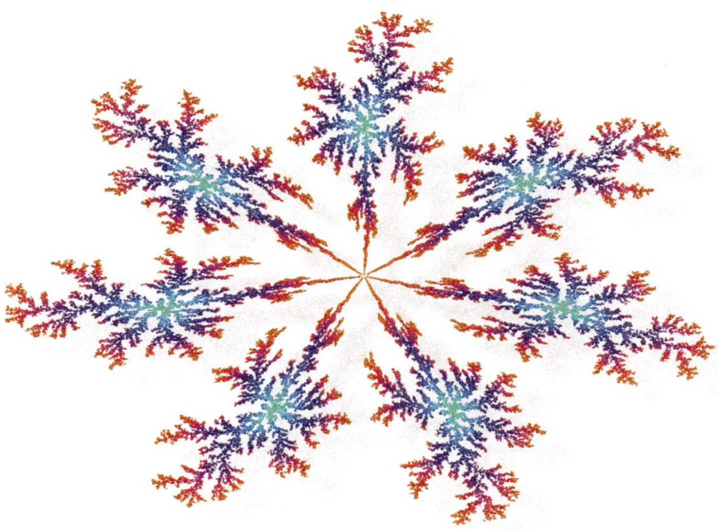

Fig. 8.11   WBRMWVWZYLABM7 $(PF = 0.134, CP = 0.119, SK = -0.844)$.

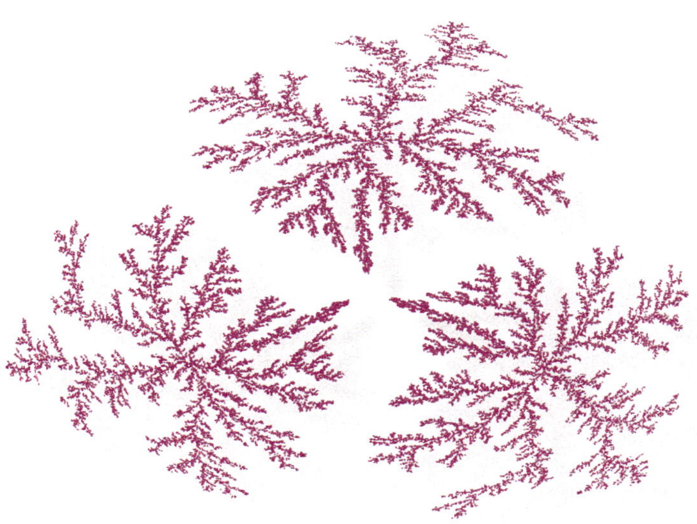

Fig. 8.12   WBTPVCBTLMIVM3 $(PF = 0.101, CP = 0.024, SK = -0.819)$.

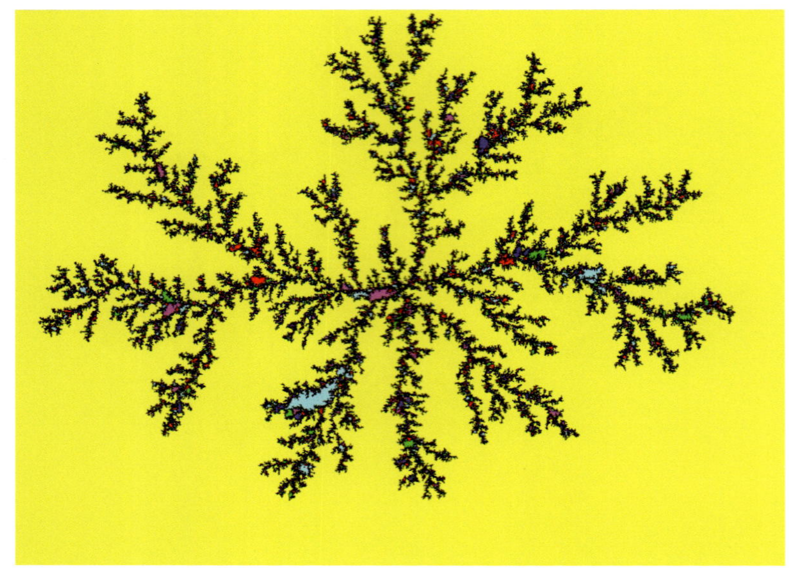

Fig. 8.13   WCBXOOFPWKOAM $(PF = 0.121, CP = 0.119, SK = -0.844)$.

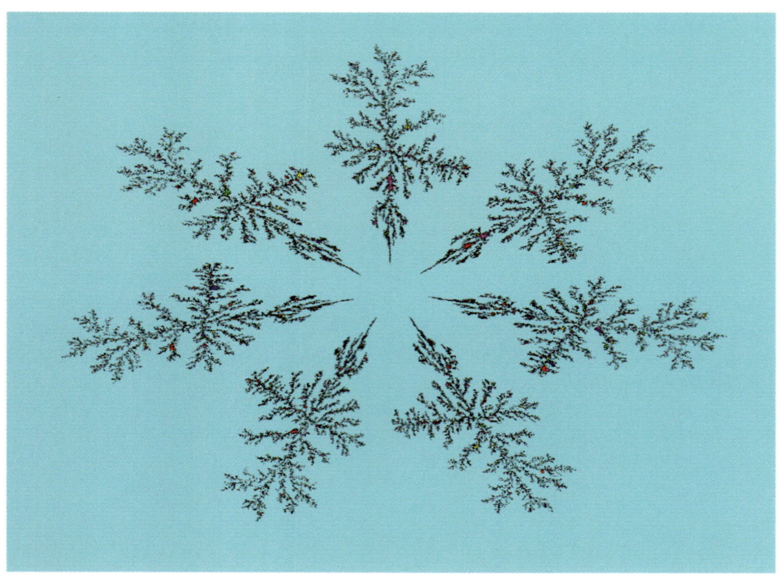

Fig. 8.14   WDPEJPGKGGDCM7 $(PF = 0.054, CP = 0.316, SK = 0.000)$.

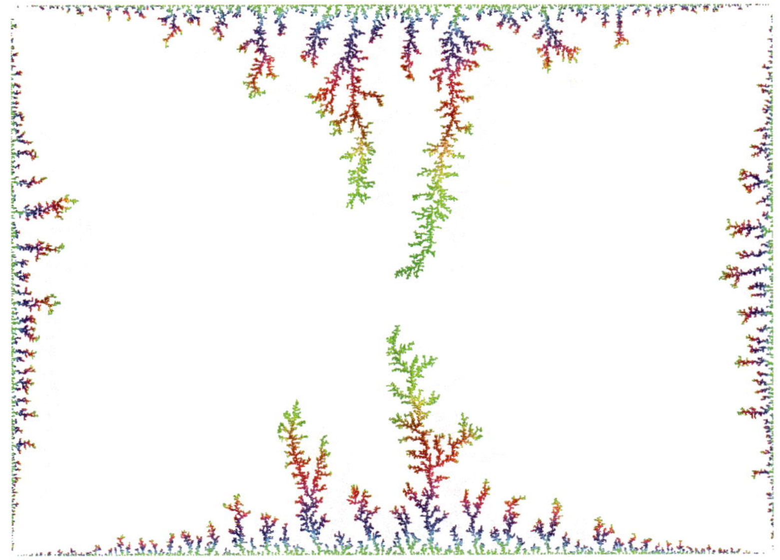

Fig. 8.15   **WAPOFBECRGTUC** $(PF = 0.075, CP = 0.033, SK = -0.009)$.

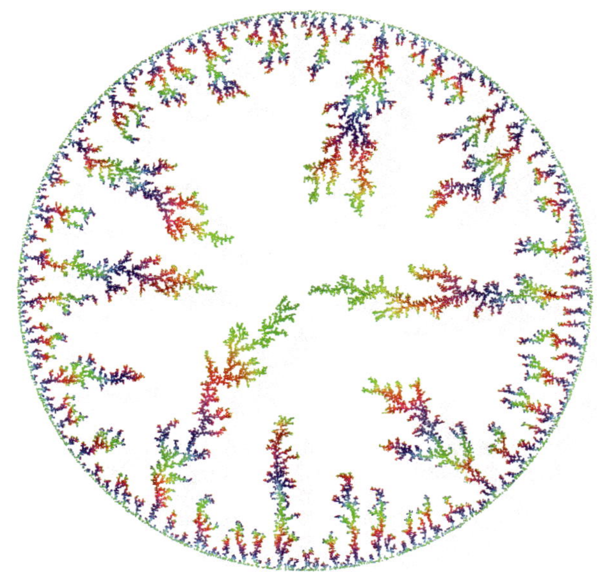

Fig. 8.16   **WAENSXLASFBKW** $(PF = 0.104, CP = 0.057, SK = -0.003)$.

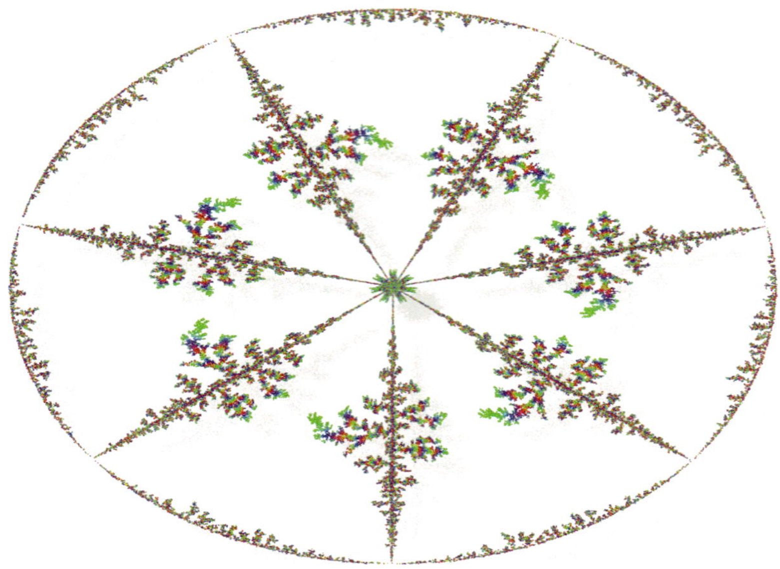

Fig. 8.17   WBBLKMKTSSZAC7 ($PF = 0.125, CP = 0.174, SK = -0.019$).

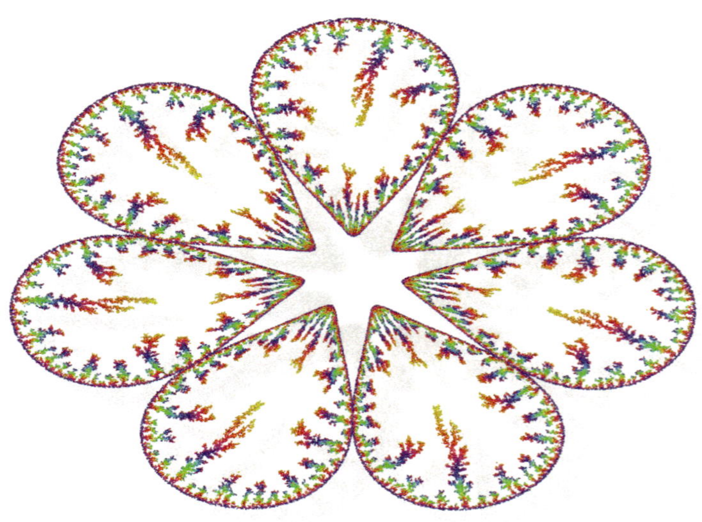

Fig. 8.18   WBLFKQLWCKFAW7 ($PF = 0.148, CP = 0.172, SK = -0.009$).

that they are icons with the corresponding number of segments. Perhaps they could be marketed as chinaware patterns.

You could also start the particles at random points throughout the region or allow them to rain down from the top but with a horizontal random walk. You could give the particles a finite size and let the size shrink as the cluster grows. You could use a Moore neighborhood rather than a von Neumann neighborhood and a neighborhood size greater than $r = 1$. You could grow three-dimensional clusters or even clusters in dimensions higher than three where other coloring options become available. DLA clusters have been simulated in dimensions as high as eight [Ball *et al.* (1984)]. These and other variants will be left for you to explore.

8.6 Stochastic Cellular Automata

The previous chapter described one-dimensional and two-dimensional cellular automata governed by deterministic rules. Cellular automata can also use random rules, and they are called *stochastic cellular automata.* The term 'stochastic' is just a fancy word for 'random,' favored by statisticians and early Russian dynamicists who often used it for what we now call 'chaos.'

There are many ways that randomness could be introduced into the cellular automaton rules, but we will consider a particular two-dimensional example called the *voter model* [Liggett (1985, 1999)] because it was originally used to describe the spatial pattern of voters in political elections [Fernández-Gracia *et al.* (2014)], but it has also been used to model invasive species [Clifford and Sudbury (1973)] and forest landscapes [Sprott *et al.* (2002)].

There are a number of variants of the voter model, but we will use a particular form with the simple rule that in each election, every voter votes the way a randomly chosen neighbor voted in the previous election. Implemented as a cellular automaton, we imagine each voter living alone in a cell within a rectangular grid of cells and that the cells are updated in a random sequence and at random times.

The only parameter in the model is the size of the neighborhood, which is taken as the area of a circle of radius r centered on each cell. Thus $r = 1$ gives four neighbors (north, south, east, and west), and $r = \sqrt{2}$ includes the four additional neighbors on the diagonals (NE, NW, SW, and SE), and so forth. In the limit of large r, the number of neighbors is given approximately by πr^2 (the area of the circle). All the neighbors in the circle are treated equally, and a random neighbor is chosen at each time step. As usual, the boundary conditions are taken as periodic, both horizontally and vertically.

Various initial conditions are possible including the extremes of completely random and completely ordered. It turns out that the patterns that result after running the model for a while are independent of the initial conditions. The former case gives *self-organization*, and the latter case gives *self-disorganization*, and so we will use random initial conditions since that is usually closer to the final state and hence requires less computation. In the limit of infinite time, the spatial pattern will disappear, and one species will take over the entire landscape, but we will not run the model nearly long enough for that to happen, but rather only for a time sufficient for each cell to be replaced a few dozen times on average.

As with the deterministic cellular automata, the cells can take on any number of discrete values, and we will denote the different values using colors as determined from Eq. (6.1). In addition, some fraction of the cells are assumed to be empty and are colored white. The code for the voter model will be prefaced by X, with the usual twelve parameters used as follows:

Case X:

Parameter	Used for
a_1	Not used, set to 0 (M)
a_2 to a_8	Seed for random number generator
a_9	Number of iterations $= 10^3 wh(a_9 + 1.3)$
a_{10}	Neighborhood size, $r = a_{10} + 2.3$
a_{11}	Number of colors $= 10a_{11} + 14$
a_{12}	Fraction of occupied cells $= a_{12} + 1.3$.

Figures 8.19 and 8.20 show two stochastic cellular automaton images produced as described above. Figure 8.19 has three colors (red, green, and blue) and a neighborhood of $r = 1.3$, and Fig. 8.20 has 22 colors and a neighborhood of $r = 3.6$, both with all cells occupied.

These images take several hours each to produce, largely because of the way the program is structured, but that is not a serious limitation because they all look rather similar, and thus you will not need to make many cases or use fancy selection schemes. The pixel fraction is entirely determined by the value of a_{12} and is given by $PF \approx \min(1, a_{12} + 1.3)$.

The cluster probability is more useful as a measure of the degree of self-organization. In fact, that is the purpose for which it was originally designed [Sprott *et al.* (2002)]. It depends strongly on the number of colors, the number of iterations, and the neighborhood size. Small neighborhood sizes tend to give more self-organization and higher values of CP. As usual,

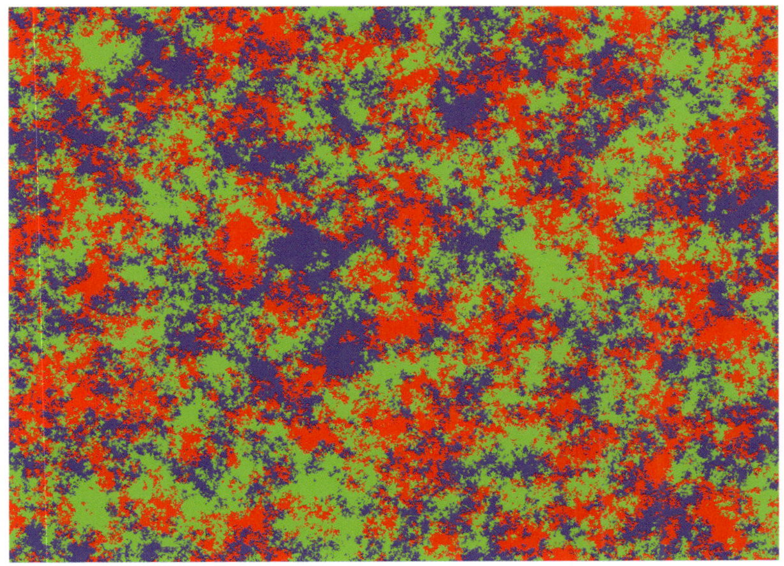

Fig. 8.19   `XMQMSFQZOMCBR` $(PF = 1.000, CP = 0.486)$.

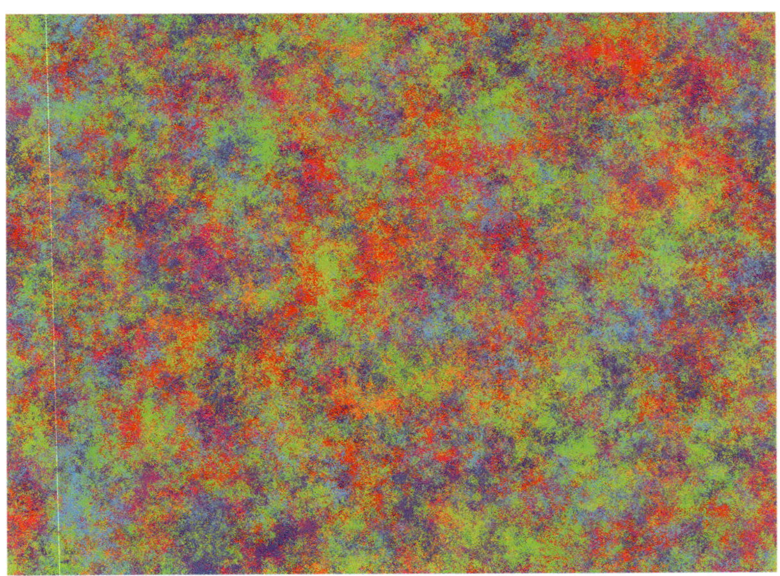

Fig. 8.20   `XMNWGAYRXGZUN` $(PF = 1.000, CP = 0.018)$.

you will probably find those cases with $CP \approx 1/3$ to be the most elegant.

There are many ways to extend the voter model. You could allow the voters to mutually influence one another, interchanging their state whenever they interact. You could use a majority rule in which they change state to that of the majority of their neighbors, which tends to eradicate the spatial structure after a while. You could choose neighbors with a probability that decreases with their distance of separation, according to a chosen *kernel function*. You could allow very distant sites to be occasionally chosen, as if the voters are strongly influenced by someone with whom they interact only electronically in what is called a *small-world network* [Watts and Strogatz (1998)]. Probably you can think of other interesting modifications of the model. While these modifications may make the model more realistic, they do not much affect the appearance or elegance of the patterns that are produced, and thus they are not especially useful or beneficial for our purpose.

8.7   Percolation

A different type of random fractal can be produced in a process called *percolation*. The idea was introduced by Broadbent and Hammersley (1957) and has been extensively studied ever since [Saberi (2015); Herega (2015)].

Consider a rectangular grid of cells in which each cell is occupied with some probability p, with the remaining cells empty. A practical example might be a forest in which some cells have trees, and others do not [Gardner *et al.* (1987)]. Now imagine that a fire starts on some boundary of the region. Clearly if the density of trees is small ($p \ll 1$), the fire will not spread, but if the density is large ($p \approx 1$) it will spread across the entire landscape leaving only isolated patches unburned.

There is a critical value of p called the *percolation threshold* (p_c) where the probability of unlimited spreading of the fire abruptly transitions from near zero to near unity. Near the percolation threshold, the burned region exhibits a fractal pattern, containing holes of arbitrarily small size (down to the limit of a single cell). Although no precise theoretical value exists for the percolation threshold, numerical experiments on a large square grid give a value of $p_c \approx 0.592745$ assuming a von Neumann neighborhood with $r = 1$ [Ziff and Sapoval (1986)]. For eight neighbors (a Moore neighborhood with $r = 1$), the value drops to $p_c \approx 0.407245$. The same model could represent the spread of disease [Kesten (1987)], the flow of a liquid through a porous medium [Hammersley (1983); Stauffer (1985)], the electrical conductivity of a mixture of dielectric and metallic components [Efros and Shklovskii (1976)], or many other similar processes [Sahimi (1994)].

To implement the scheme numerically, we use an initial condition in which cells are colored in a chosen (red, grn, blu) color with a probability $1 - p$. Then we use the PAINT command that is available in BASIC to color the white regions in one of six different colors (the 'coloring book scheme') starting from every white point on the top row. Values are chosen for p in the range of $0.576 \leq p \leq 0.626$ to encompass the percolation threshold as given by the following:

Case **Y**:

Parameter	Used for
a_1	Not used, set to 0 (M)
a_2 to a_8	Seed for random number generator
a_9	Red intensity, $red = 120 + 100a_9$
a_{10}	Green intensity, $grn = 120 + 100a_{10}$
a_{11}	Blue intensity, $blu = 120 + 100a_{11}$
a_{12}	Fraction of empty cells, $p = 0.6 + 0.02a_{12}$.

Images produced in this way are shown in Figs. 8.21 and 8.22. Figure 8.21 is a case that is barely below the percolation threshold ($p = 0.588$) where the clusters die just before reaching the bottom of the plot. By contrast, Fig. 8.22 is a case that is barely above the percolation threshold ($p = 0.590$), and the large cluster is filled with a fractal distribution of holes. This is an example of a *fat fractal* [Umberger and Farmer (1985)] since it has an integer dimension of 2.0 and a finite area but with infinitely many holes down to arbitrarily small size.

It would perhaps be better and more natural to use periodic boundary conditions (the right edge of the region is adjacent to the left edge), but that is not how the BASIC PAINT command works, and thus the reward for doing that is not worth the effort. It will be left as a challenge for you to emulate the PAINT command with periodic boundaries in your favorite programming language if it does not include a similar functionality.

You can automate the selection of elegant cases using the pixel fraction. If the percolation front dies quickly, then PF will have a value close to $1 - p$, or about 0.4 representing only the *dangling clusters*, whereas if it fills the entire space with few holes, then PF will approach 1.0. The elegant cases have intermediate values in the range of $0.6 \lesssim PF \lesssim 0.75$, and that filter will eliminate about three quarters of the cases that are not likely to be interesting. The cluster probability is less useful since nearly all the cases are in a narrow range of about $0.07 \lesssim CP \lesssim 0.10$.

Finally, Fig. 8.23 shows a larger, higher-resolution example of a percolation fractal near the percolation threshold.

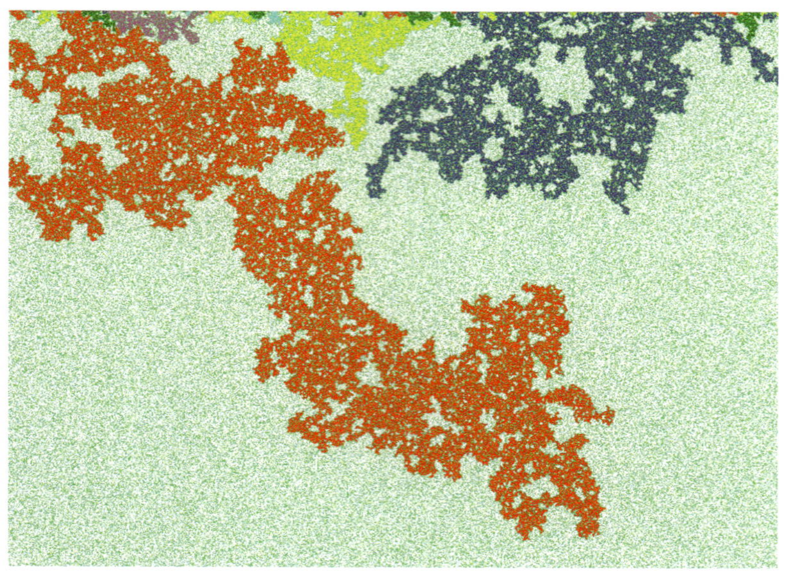

Fig. 8.21　YMHQZXISBFWNG $(PF = 0.590, CP = 0.083)$.

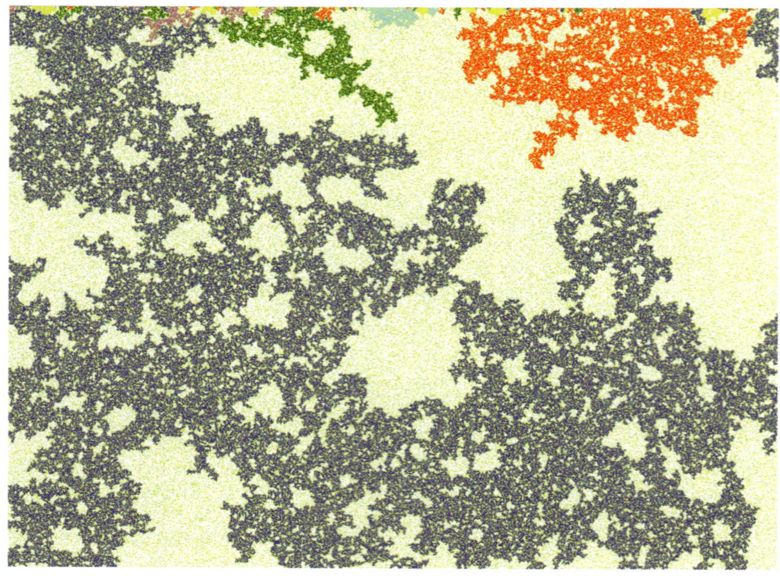

Fig. 8.22　YMGWQBQHNWXLH $(PF = 0.730, CP = 0.083)$.

Fig. 8.23   YMNBUWVESIXSL ($PF = 0.836, CP = 0.087$).

Chapter 9

Fractal Tessellations

The final example of fractals will involve tilings of the plane. Often such patterns are tediously repetitious, but we will consider examples that are aperiodic and that have structure down to the smallest visible scale. Some of the examples are deterministic, while others are random and thus represent an extension of the cases in the previous chapter.

9.1 Regular Tilings of the Plane

Tessellation is a fancy word that refers to completely filling a space with objects without any gaps or overlaps and usually subject to additional rules such as the symmetry, types, number, and shapes of the objects. Due to the limitations of the printed page and computer screen, we will consider only two-dimensional tessellations of the plane, although the ideas can be extended to non-Euclidean geometries and to arbitrarily high dimensions, of which the honeycomb is the most familiar three-dimensional example [Nazzi (2016)]. Tessellations have been extensively studied by mathematicians [Grünbaum and Shephard (1987)] and are often employed by architects and artists, perhaps the best known of whom is the Dutch graphic artist M. C. Escher (1958).

The simplest example of a tessellation involves filling the plane with identical shapes, called a *monohedral* tiling. If the tiles are regular polygons, it turns out that only three such tilings are possible [Coxeter (1973)], using equilateral triangles (internal angles of 60°), squares (internal angles of 90°), and hexagons (internal angles of 120°) as shown in Fig. 9.1. The reason is that these are the only regular polygons whose internal angle is a divisor of 360°. For example, the regular pentagon cannot tile the plane because

205

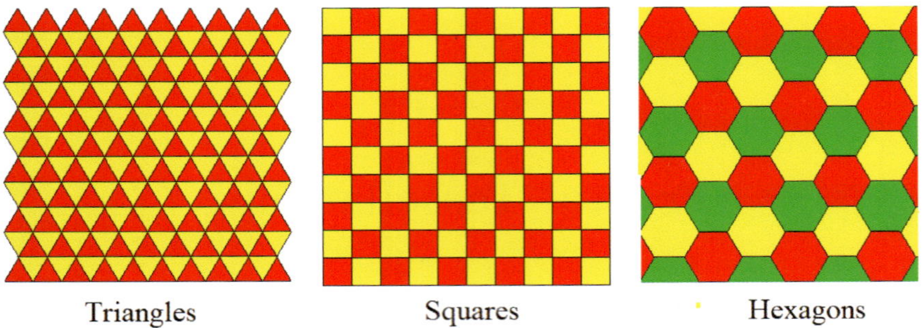

Triangles   Squares   Hexagons

Fig. 9.1   The three regular tilings of the plane (using regular polygons).

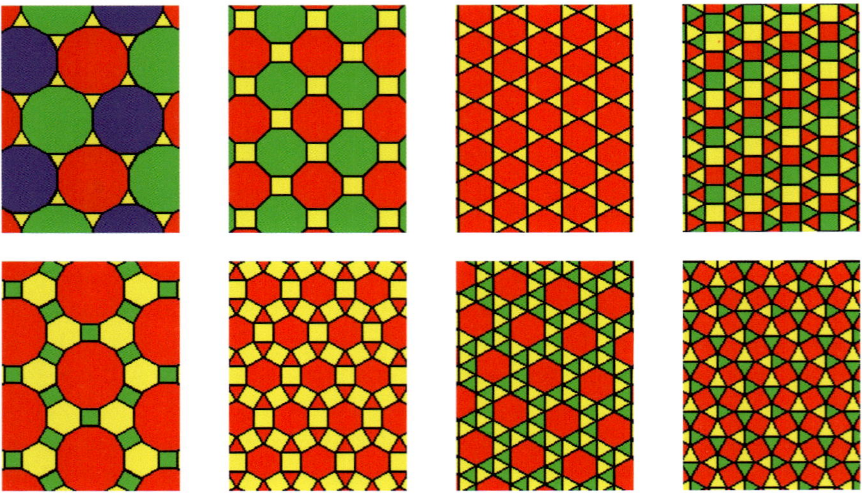

Fig. 9.2   The eight semiregular tilings of the plane (using two or three regular polygons).

it has internal angles of 108°, and so three of them placed around a circle would leave a gap of 36° whereas four of them would overlap by 72°.

These regular tilings are neither fractals nor are they especially elegant, even with added color. Appel and Haken (1989) provided a surprisingly difficult proof of the *four-color theorem* which states that four colors are sufficient for *any* tiling of the plane where the regions that share a common boundary line (called an *edge*) have different colors as a cartographer might desire. In fact, for the regular tilings in Fig. 9.1, only two colors are required for the triangles and squares, and three colors for the hexagons.

If you relax the constraints that all the polygons must be identical but require that the pattern is otherwise repeating, there are eight additional possible tilings that are called *semiregular* (or *Archimedean* or *uniform*) as shown in Fig. 9.2. They each consist of either two or three regular polygons.

In addition to being regular (or semiregular), these tilings are *periodic*. Seeing a sufficient portion of one allows you to fill the entire plane correctly by translating it in two different directions since the pattern repeats exactly. Wallpapers are often printed in this way. In fact, such periodic tilings are called *wallpaper groups*, of which seventeen types exist [Armstrong (1988)].

9.2 Penrose Tilings

Most tilings contain irregular shapes, a familiar example of which is a jigsaw puzzle. A trivial example of an irregular tessellation involves stretching of one of the regular patterns in Fig. 9.1 so that the triangles are no longer equilateral or the squares become rectangles. You can also shear those patterns so that, for example, the squares become *rhombuses*, or more generally, *parallelograms*. In fact, any triangle or quadrilateral can be used to tile the plane since it can be placed adjacent to its mirror image to form a parallelogram. As a reminder, a rhombus is a four-sided figure with all sides of equal length and hence with opposite sides parallel and opposite angles equal, and a parallelogram is similar except that adjacent sides can be of different lengths.

Clearly, the variety of possible irregular tilings is enormous, but further discussion of this fascinating topic would take us too far afield. Instead, we will consider a particular example of an irregular tiling that consists of only two rhombuses and is aperiodic with fractal properties. Figure 9.3 shows an example of such a *Penrose tiling*, named after the mathematician and physicist Sir Roger Penrose (1974).

This figure is only one of three known types of Penrose tilings (called P1, P2, and P3), but P3 is in some sense the simplest. It looks like it consists of numerous small cubes arranged at odd angles, but if you look more closely, you will see that it contains only two rhombuses, one with interior angles of $36°$ and $144°$ (called the *thin rhomb*), and the other with internal angles of $72°$ and $108°$ (called the *thick rhomb*). They can be colored using only three colors as shown in the figure. The Penrose

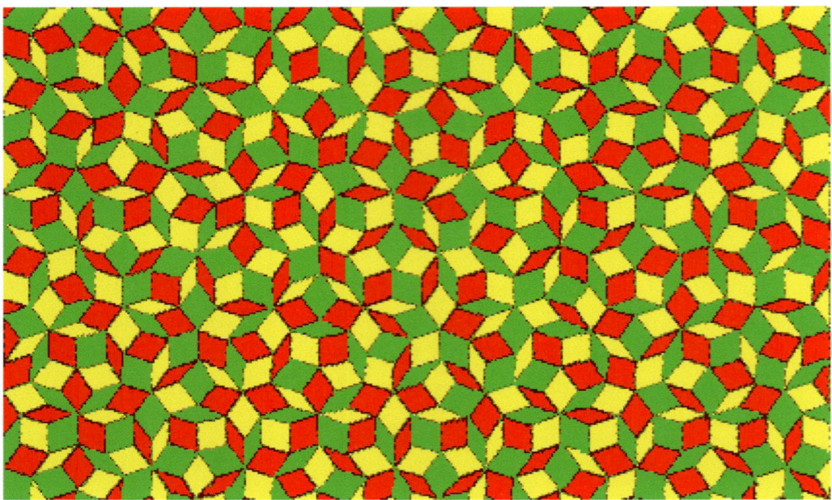

Fig. 9.3   A Penrose (aperiodic) tiling of the plane.

tilings have a number of other curious features:

- They are *aperiodic* in the sense that no translation of the pattern will leave it unchanged.
- They can be constructed with five-fold rotational symmetry.
- Every finite region occurs infinitely many times, but not in a periodic manner.
- Patterns that occur on one scale reoccur at successively larger and smaller scales (called *inflation* and *deflation*), giving the object a self-similar fractal quality.

9.3   Cantor Squares

The *triadic Cantor set* (or *Cantor ternary set*) [Peitgen *et al.* (2004)] is one in which the middle third of a finite line is removed, and then the middle third of the two remaining lines are removed, and the process is repeated forever, or more practically, until the remaining segments of the line are single pixels in width. The idea can be generalized to a two-dimensional square or rectangular region, producing what is called a *Cantor square* or

Sierpiński carpet [Sierpiński (1916)]. Of course there is nothing special about the factor of one third, and so you can produce a variety of Cantor squares with varying fractal dimensions and colors. The 'coloring book scheme' using the BASIC PAINT command is a convenient way to add color. Two examples produced in this way are shown in Figs. 9.4 and 9.5 with parameters used as follows:

Case ZA:

Parameter	Used for
a_1	Set to -1.2 (A)
a_2 to a_8	Seed for random number generator
a_9	Contraction factor, $f = (13 + 10a_9)/54$
a_{10} to a_{12}	Not used, set to 0 (M).

Since we have previously used all the letters of the alphabet as codes except Z, the cases in this chapter will use a second letter (A in this case) to denote the type of tessellation. The random numbers are used here only to determine the choice of colors from the six available. The contraction factor f would be $1/3$ for the triadic Cantor square, and that case is shown in Fig. 9.4. The allowed values of f span the range from $1/54$ to $13/27$. Values of f larger than $1/2$ would fill the entire region with black. Figure 9.5 shows a case with less contraction ($f = 6/13$). The backbone Cantor set shown in black is produced using iterated function systems with the random iteration algorithm as described in Chapter 5.

9.4 Stochastic Fractal Squares

The previous examples have been deterministic, with the random numbers used only to choose the colors and to advance the iterated function system with a fixed contraction factor. You can also allow the contraction factor to be chosen randomly in the range of 0 to 0.5 at each iteration. The distribution of sizes is determined by the number of such lines that are drawn, and so we use that as a parameter of the system in the range of $10 \leq a_9 \leq 260$ according to the following:

Cases ZB to ZG:

Parameter	Used for
a_1	Set to -1.1 (B) to -0.6 (G)
a_2 to a_8	Seed for random number generator
a_9	Number of iterations $= 130 + 100a_9$
a_{10} to a_{12}	Not used, set to 0 (M).

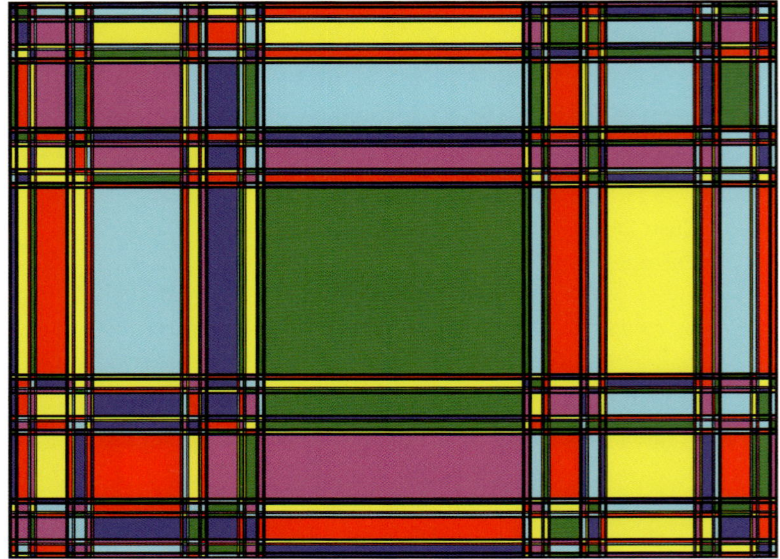

Fig. 9.4   **ZAAUJOGVTRMMM** $(PF = 0.273, CP = 0.733)$.

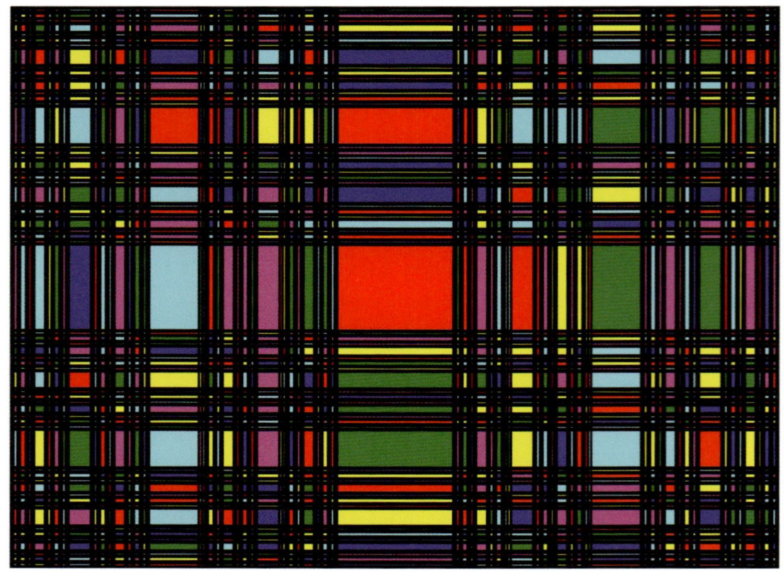

Fig. 9.5   **ZAIYVQGVWWMMM** $(PF = 0.680, CP = 0.752)$.

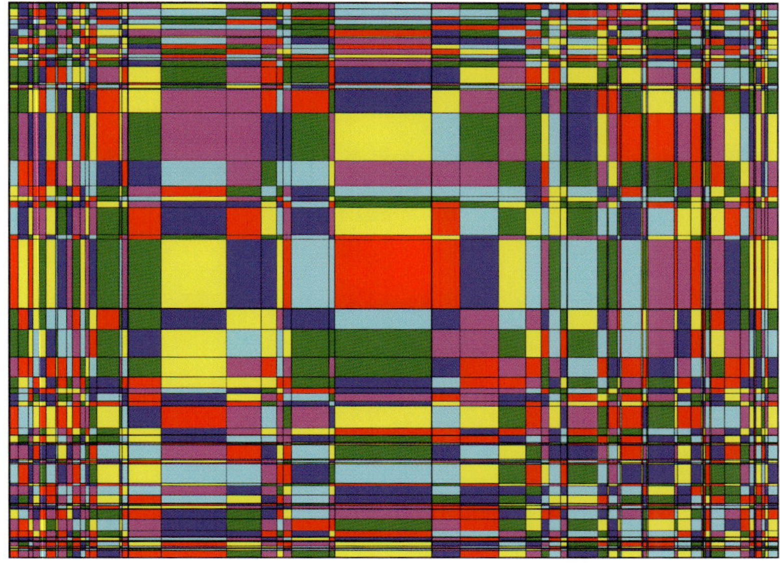

Fig. 9.6 ZBIYDZNZHHMMM ($PF = 0.128, CP = 0.696$).

A typical example of such a system is shown in Fig. 9.6. This case has 80 horizonal lines and 80 vertical lines, and thus $81 \times 81 = 6561$ rectangles, most of which are invisibly small.

If you look carefully, you will note that the patterns have a certain symmetry (except for the colors) that results from using the same sequence of contractions and iterations horizontally and vertically. It is easy to remove that restriction to produce patterns like the one in Fig. 9.7, which was chosen to have the same number of lines as in Fig. 9.6 but without the symmetry. Which case would you consider most elegant? Perhaps the difference is not even noticeable to you.

An even simpler method is to draw the horizontal and vertical lines at random positions. Figure 9.8 shows such an example, again chosen to have 80 lines in each direction to facilitate comparison. It has the same number of rectangles in a wide variety of sizes but with a distribution that is in some sense 'less fractal' than the previous cases.

Rather than drawing horizontal and vertical lines, you could instead begin at random points in the region and from each point draw a line at an angle chosen randomly in the range of 0 to 180° (0 to π radians). The

 Elegant Fractals

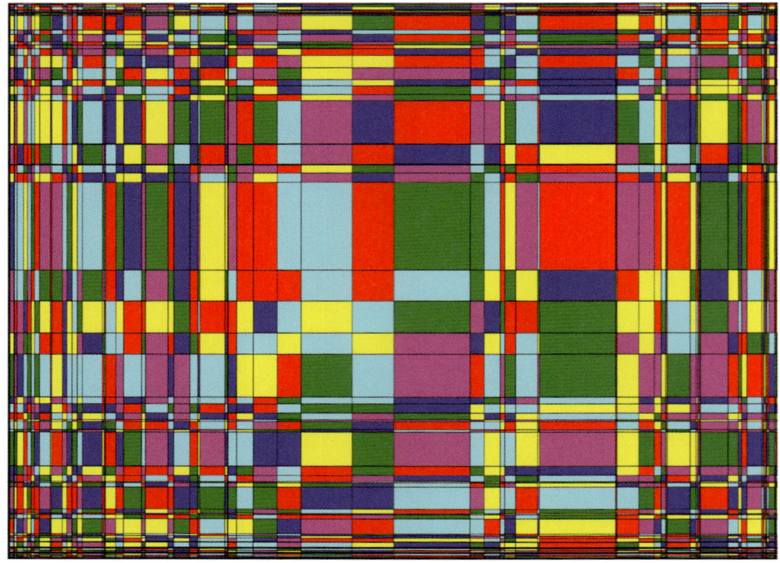

Fig. 9.7   ZCMGHIVDSHMMM $(PF = 0.125, CP = 0.704)$.

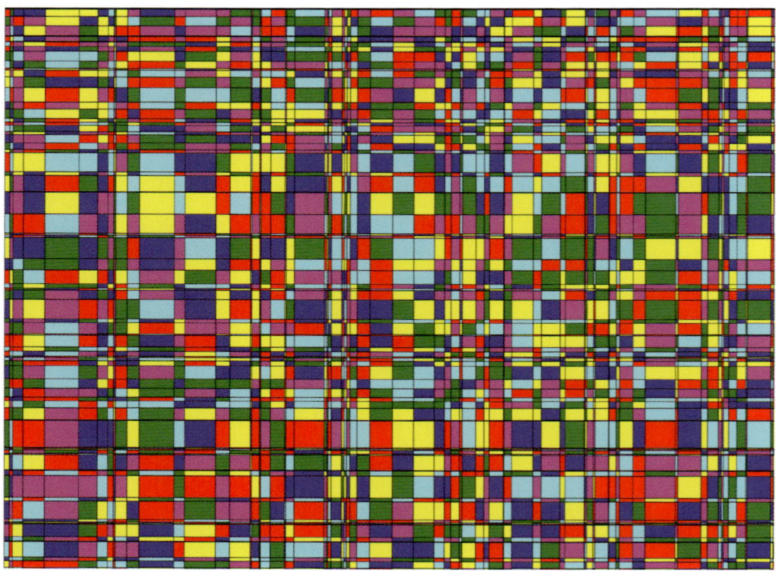

Fig. 9.8   ZDIVSBUXXHMMM $(PF = 0.143, CP = 0.642)$.

result is to tile the plane in irregular *convex polygons*, mostly triangles and quadrilaterals, but with some higher-order irregular polygons. Figure 9.9 shows an example with 80 such lines.

9.5 The Gilbert Tessellation

In all the previous examples, the polygons share *vertices*, which means that nowhere do three polygons meet at a 'T-junction.' That limitation is relaxed in the *Gilbert tessellation* [Gilbert (1967)] or *random crack network* [Gray *et al.* (1976)] in which cracks begin at random points and move in both directions along a straight line at a random angle until they encounter another crack (line). The cracks can form one at a time or simultaneously, the former being easier to program and the latter being a better model of mudcrack patterns in nature [Gray *et al.* (1976)]. An example of the first case with 220 lines is shown in Fig. 9.10.

There is a variant of the model in which the cracks are restricted to move only horizontally or vertically [Mackisack and Miles (1996)], tiling the plane nonuniformly with rectangles, one example of which with 260 lines is shown in Fig. 9.11. This is a case where allowing the cracks to grow

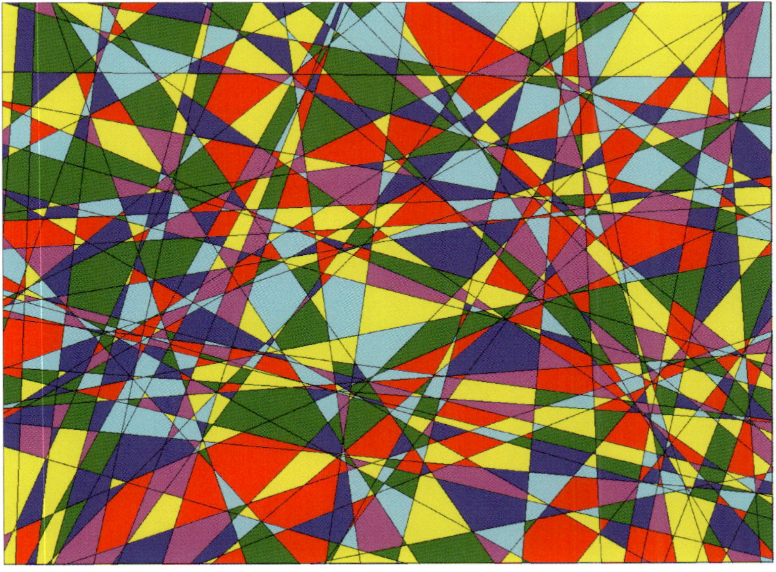

Fig. 9.9   ZETKUNWTYHMMM ($PF = 0.071, CP = 0.809$).

Elegant Fractals

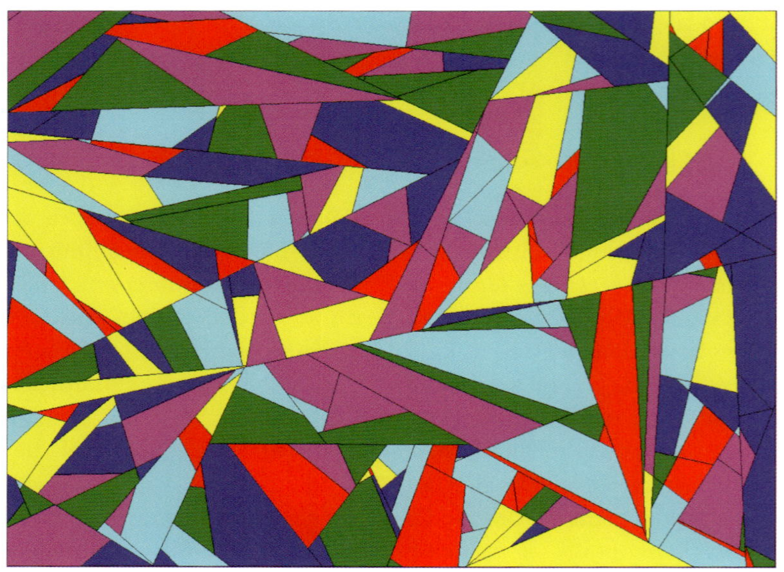

Fig. 9.10   **ZFKDWQHXWVMMM** $(PF = 0.033, CP = 0.907)$.

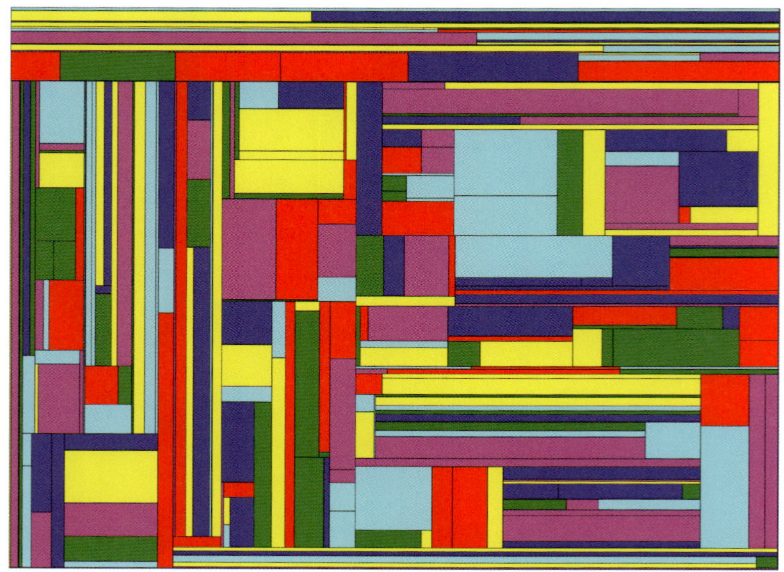

Fig. 9.11   **ZGUGCQWTTZMMM** $(PF = 0.070, CP = 0.810)$.

simultaneously reduces the number of highly elongated rectangles, although that was not done for this figure.

9.6   The Logarithmic Spiral

In all the previous tessellations, the boundaries between the regions are straight lines. Such boundaries can also be curves, a classic example of which is the *logarithmic spiral* (also called an *equiangular spiral* or *growth spiral*), which is exactly self-similar and thus rightly deserves to be called a fractal. It circles the origin infinitely many times but has a finite length.

The logarithmic spiral was first described by René Descartes and was further studied by Jacob Bernoulli who called it *spira mirabilis* ('the marvelous spiral'). It is distinct from the *Archimedean spiral* (also called the *arithmetic spiral*), such as early gramophone records, which are not self-similar. Many objects in nature are approximately logarithmic spirals including the spiral arms of galaxies [Bertin and Lin (1996)], the bands in hurricanes [Gray (1901)], and biological structures such as molluscan shells [Cortie (1992)]. Most of the spirals from the Julia sets in Chapter 6 are logarithmic spirals.

The logarithmic spiral is governed by the equations

$$
\begin{aligned}
x(t) &= ae^{bt}\cos t \\
y(t) &= ae^{bt}\sin t,
\end{aligned}
\tag{9.1}
$$

where a is the initial distance from the origin and b is the growth rate (or decay rate if negative). Since the curve is self-similar (it looks the same at any magnification), we can take $a = 1$ without loss of generality and take b negative since the sign of b only determines whether increasing t (time) corresponds to outward or inward spiraling, both producing the same pattern.

Of course a single curve spiraling into the origin trivially tiles the plane with a single space-filling tile with a slit for the spiral and produces an inelegant image. To produce a more interesting image, you can connect points along the spiral separated in time by δt with straight lines. Then the parameters are b, which determines the *pitch* ϕ of the spiral according to $\phi = 1/\arctan(1/b)$, and δt which determines the size and shape of the individual tiles along the direction of the spiral. These quantities are coded in the following way:

>   Case ZH:
>   Parameter   Used for
> a_1 Set to -0.5 (H)
> a_2 to a_8 Seed for random number generator
> a_9 Parameter $b = -(1.3 + a_9)/5.4$
> a_{10} Parameter $\delta t = 3.9 + 3a_{10}$
> a_{11} and a_{12} Not used, set to 0 (M).

Figures 9.12 and 9.13 show two examples of logarithmic spirals obtained in this way. The former case has a relatively small value of $\delta t = 1.5$ radians, and the latter has a larger value of $\delta t = 3.0$ radians. Figure 9.12 does not tile the entire plane because of the white space, but Fig. 9.13 does if allowed to continue spiraling outward.

These objects are fractals in the sense that zooming in on the origin reveals a pattern that is self-similar to the whole, although with different colors, and they contain tiles that are self-similar with arbitrarily small sizes.

9.7   Apollonian Circles

The *Apollonius problem* [Coxeter (1968)] is to find a circle that is simultaneously tangent to three other circles. It is named after the Greek geometer and astronomer Apollonius of Perga who lived from circa 262 BC to 190 BC. Such circles are called *circles of Apollonius*.

Starting with three arbitrary (not necessarily identical) non-overlapping circles that mutually touch, one can fill in the curvilinear triangle between them with an Apollonian circle, creating three smaller curvilinear triangles that can be filled with three smaller Apollonian circles, and the process repeated until all that remains is a gasket with zero area and a fractal dimension of approximately 1.3 [Mauldin and Urbanski (1998)], one symmetric example of which is shown in Fig. 9.14.

The circular regions of the figure that are not part of the gasket are shown in colors and can be considered as a fractal tessellation in which there are infinitely many successively smaller circular tiles. The pattern can be extended to cover an area of arbitrary size by making three such circular regions enclosed by an even larger circle and arranging them in Apollonian form, and continuing the process until the entire plane is filled.

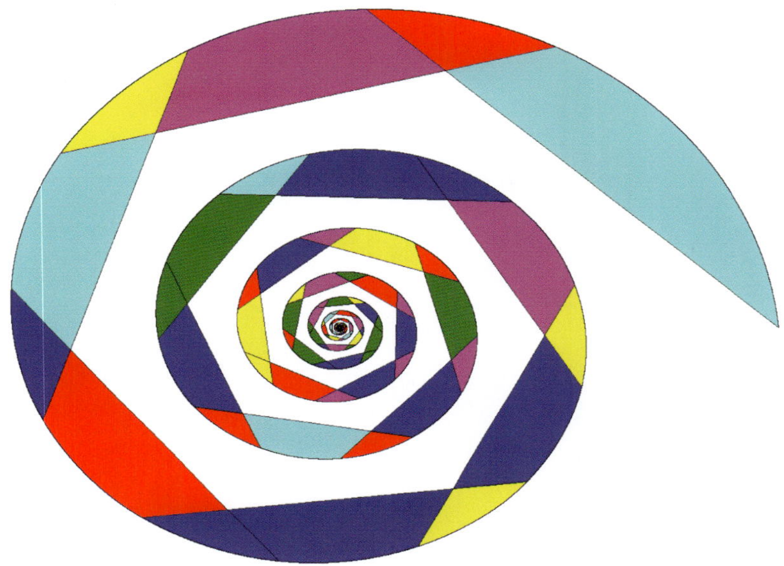

Fig. 9.12   **ZHWVDQOOYEEMM** ($PF = 0.016, CP = 0.959$).

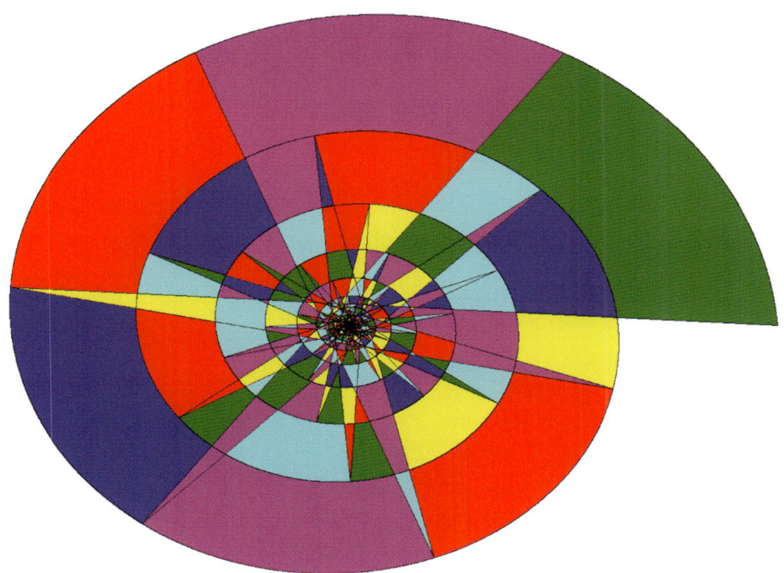

Fig. 9.13   **ZHQSAPAZVDJMM** ($PF = 0.021, CP = 0.948$).

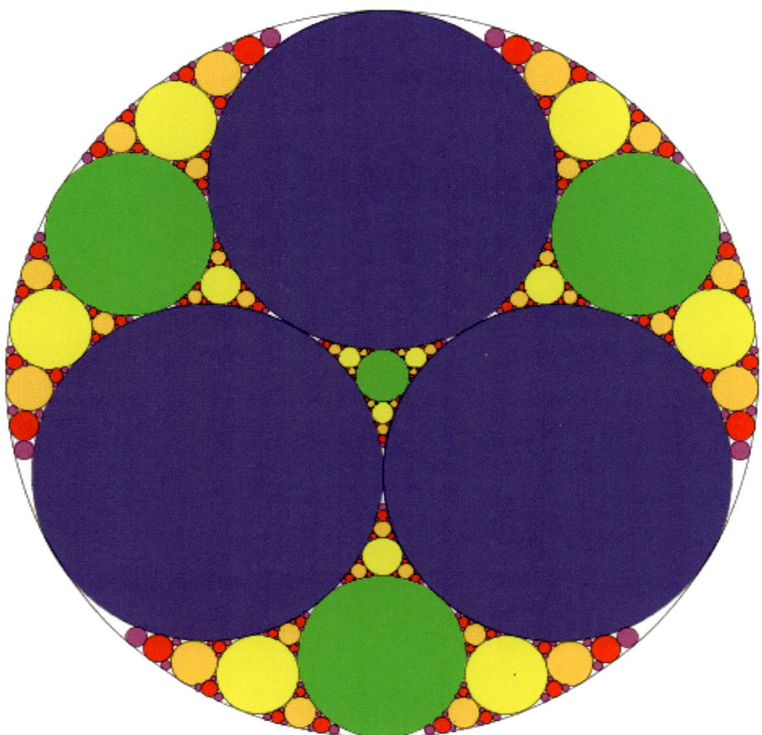

Fig. 9.14   Apollonian circles.

More simply, the image can just be enlarged as necessary to fill the desired space, although the three large blue circles may then become undesirably large.

9.8   Shier Tilings

Apollonian gaskets have a symmetry that makes them attractive but that also limits their variety. Thus we now consider tessellations that are fractal but that also have a random component. These ideas have been recently developed by John Shier [Shier and Bourke (2013)], and they are related to the practical problem of the optimal packing of objects into a finite region of space for which there is a large mathematical literature [Lodi *et al.* (2002); Dodds and Weitz (2002)].

Consider first the case of ellipses whose areas obey a power law but that are placed randomly in the plane without regard to overlap except that they must fit within a rectangular boundary. The aspect ratio of each ellipse is the same as the aspect ratio of the bounding rectangle so that they would be circles if the bounding rectangle were a square, and they are all oriented in the same direction. The power law is chosen in such a way that the total area of the ellipses is the same as the area of the bounding rectangle in the limit of infinitely many of them in anticipation of later using them to tile the plane.

If the bounding region has area A, then successive ellipses have areas given by

$$A_i = \frac{A}{\zeta(c, N)(N + i)^c} \qquad (9.2)$$

for $i = 0, 1, 2, \ldots$ where $\zeta(c, N)$ is the Hurwitz zeta function [Vepstas (2007)] defined by

$$\zeta(c, N) = \sum_{m=0}^{\infty} \frac{1}{(N + m)^c}. \qquad (9.3)$$

The parameter $c > 1$ is the power law exponent, and $N > 0$ determines the area A_0 of the first ellipse according to $A_0 = A/\zeta(c, N)N^c$. For $N = 1$, the Hurwitz zeta function reduces to the more familiar Riemann zeta function $\zeta(c, 1)$ in which the variables are usually taken to be complex numbers. Equation (9.3) is just one of many generalizations of the Riemann zeta function [Ivic (1985); Titchmarsh (1986); Karatsuba and Voronin (1992)]. You should convince yourself that $\sum_{i=0}^{\infty} A_i = A$ so that they would completely tile the plane if positioned with no overlaps.

Figure 9.15 shows an example of such an image with parameters chosen according to

Cases **ZI** to **ZZ**:

Parameter	Used for
a_1	Set to -0.4 (**I**) to 1.3 (**Z**)
a_2 to a_8	Seed for random number generator
a_9	Number of tiles $= 700 + 500a_9$
a_{10}	Parameter $N = 3.2 + a_{10}$
a_{11}	Parameter $c = 1.26 + 0.2a_{11}$
a_{12}	Not used, set to 0 (**M**).

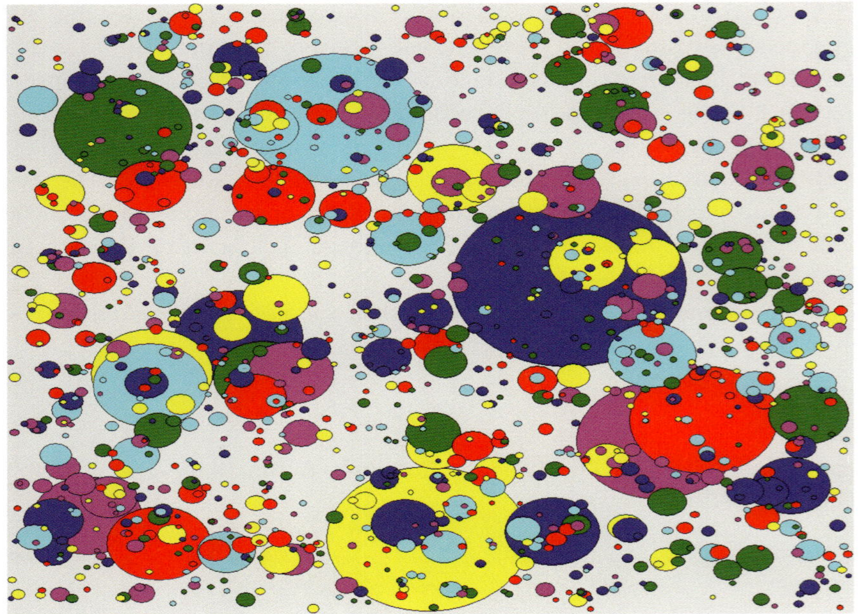

Fig. 9.15　`ZIXIRRZKWZPJM` ($PF = 0.535, CP = 0.844$).

The colors cycle through the six 'coloring book' colors. Perhaps this figure reminds you of craters on the Moon.

The idea can be extended to rectangles with the same orientation and aspect ratio as the bounding rectangle, and such an example is shown in Fig. 9.16. Of course neither of these cases are tessellations since they are overlapping and not space-filling. In particular, the light gray gasket has a nonzero area equal to the sum of the areas of the overlapping parts of the rectangles in the limit of infinitely many such rectangles. Accordingly, the pixel fraction PF is much less than 1.0 and is the order of 0.5 for these images. Perhaps they are also too unstructured to be considered elegant.

To obtain a proper tessellation, it is necessary to place the tiles so there is no overlap. It seems that this would be difficult to do, but remarkably it has been proved [Ennis (2016)] that for an appropriate range of c and N, circular tiles can be placed at random within the allowed space, and they will completely fill the space except for a gasket of measure zero (zero area) in the limit of infinitely many such tiles. Furthermore, the size and shape

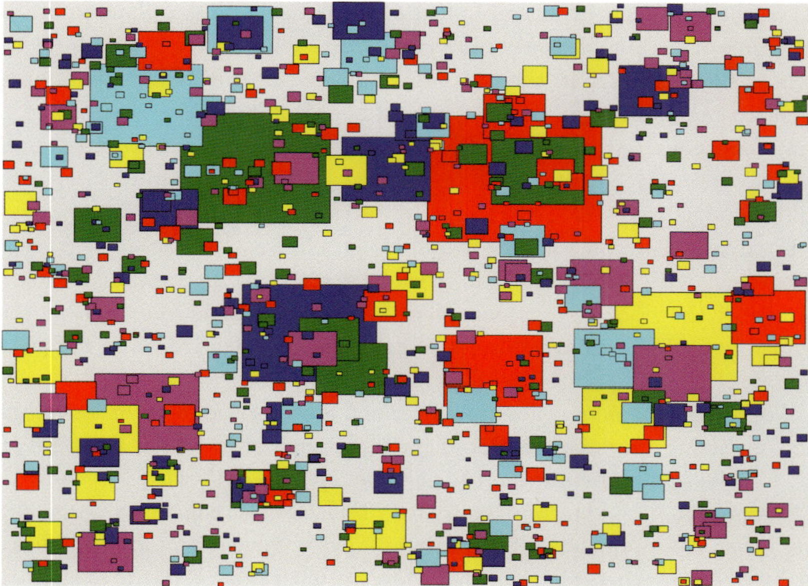

Fig. 9.16   ZJGKKFSVJZPGM ($PF = 0.494, CP = 0.813$).

of the tiles appear not to matter provided their area is scaled according to
Eq. (9.2).

To dramatically illustrate the idea, suppose you want to tile your bathroom floor in such a way. All you need is a stack of tiles with areas given
by Eq. (9.2), sorted from largest to smallest. You take the largest tile and
place it anywhere you want. Then you take the next larger tile and put it
anywhere in the remaining space. You continue the process until you run
out of tiles, always finding that there is a space available for each one until
all that remains is a space of negligible area between the individual tiles
that you can easily fill with a small amount of grout. You never need to
backtrack to move one of the previously placed tiles to make room for the
next one, and so you can cement them into place as you proceed. Curiously,
the process is irreversible since if you start with the smallest tile, you will
almost always run out of space long before using the larger tiles with large
gaps remaining.

However, this works only if c is not too large and N is not too small.
Shier recommends starting with $c = 1.25$ and $N = 2$. Otherwise, there is a

chance that the algorithm will halt, being unable to find a place to put one of the tiles or making it overly difficult to find such a place. Fortunately, this tends to happen early in the process, but it requires a 'bail-out' condition, which we take as ten million random trials. Don't let this discourage you from tiling your bathroom floor in this way since you can do much better than random placements, and presumably a better computer algorithm could do so as well.

Figures 9.17 and 9.18 show examples for ellipses and rectangles, respectively, each with 1350 tiles. In these cases, the pixel fraction PF is the fraction of the floor that is occupied by tiles, and it is less than 1.0 only because you did not have infinitely many tiles. Shier refers to PF as the *fill factor*.

Would you consider a floor tiled in such a way to be more or less elegant than one using more conventional patterns such as those in Figs. 9.1 and 9.2? How would you compare the elegance of Fig. 9.17 with Fig. 9.14?

To show that the method works for other shapes, consider regular polygons other than squares. The main difficulty for such more complicated shapes is calculating whether they overlap with one of the previous tiles. However, it is relatively simple to find the bounding circle and use that as the criterion, giving a pixel fraction somewhat less than 1.0 and a correspondingly larger gasket.

Figures 9.19 to 9.22 show the result for pentagons, hexagons, heptagons, and octagons, respectively, which form a succession approaching a circle in the limit of a polygon with infinitely many sides. In all four of these figures, the polygons are oriented identically with horizontal bottoms.

One virtue of determining overlap by considering the bounding circles is that the polygons can be rotated randomly without any additional calculation. Figures 9.23 to 9.26 show the same four polygons as in the previous figures but rotated randomly. Which version do you think is more elegant?

Of course, you can also combine the various polygons in a single image with random orientations as shown in Fig. 9.27. Do you find this more or less elegant than the ones with the same types of polygons?

An endless variety of other shapes is possible, but to simplify the overlap testing we will consider just two additional examples. Figure 9.28 shows a case where the tiles are in the form of a *Koch snowflake* (also called a *Koch island*) named after the Swedish mathematician Helge von Koch who studied such fractal types over a hundred years ago [von Koch (1904)].

The tiles are formed by starting with an octagon and replacing each of the eight sides by a *Koch curve*, which in this case has eight segments,

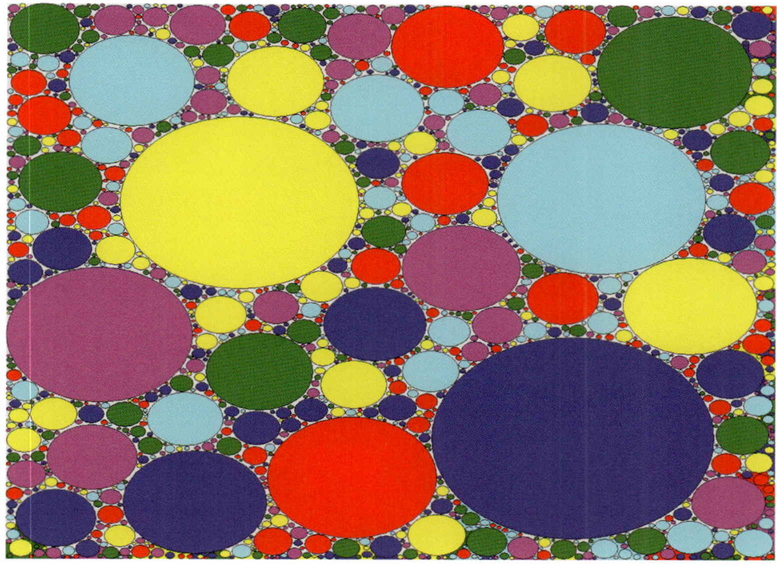

Fig. 9.17   **ZKTVDHFEJZSUM** ($PF = 0.920, CP = 0.845$).

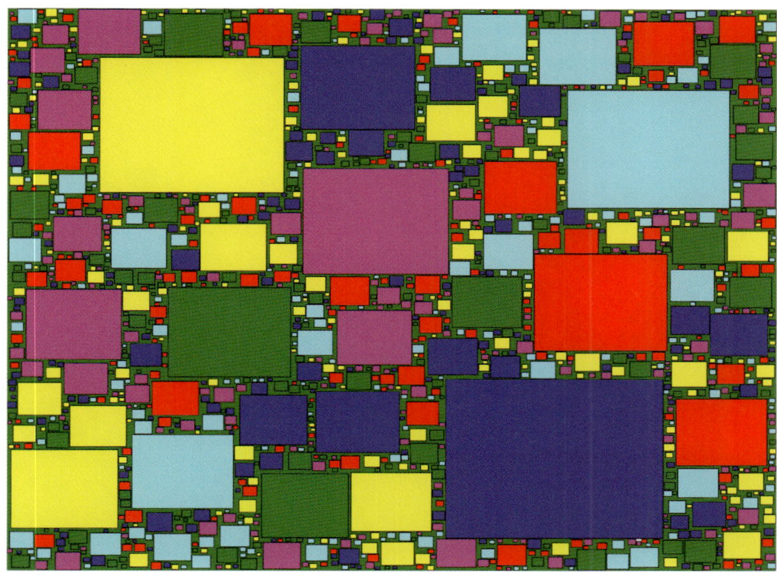

Fig. 9.18   **ZLFARFLZHZPNM** ($PF = 0.853, CP = 0.792$).

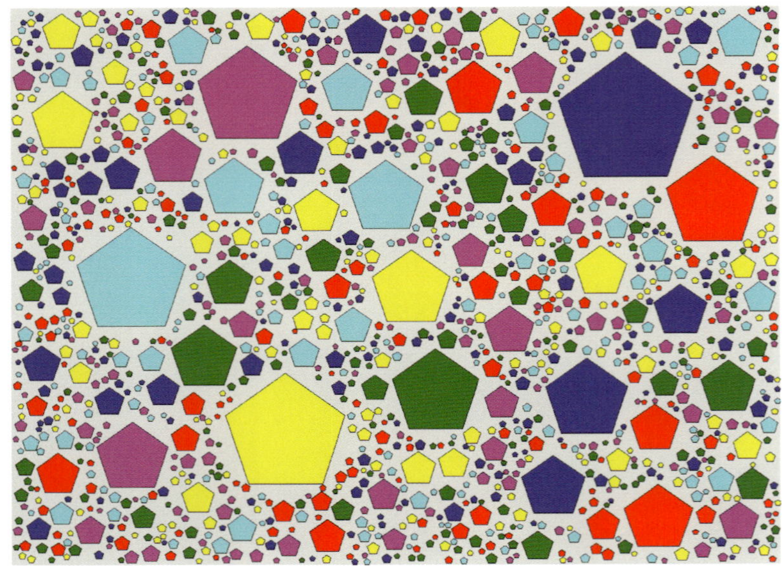

Fig. 9.19   ZMZTCFSSKZWFM $(PF = 0.517, CP = 0.820)$.

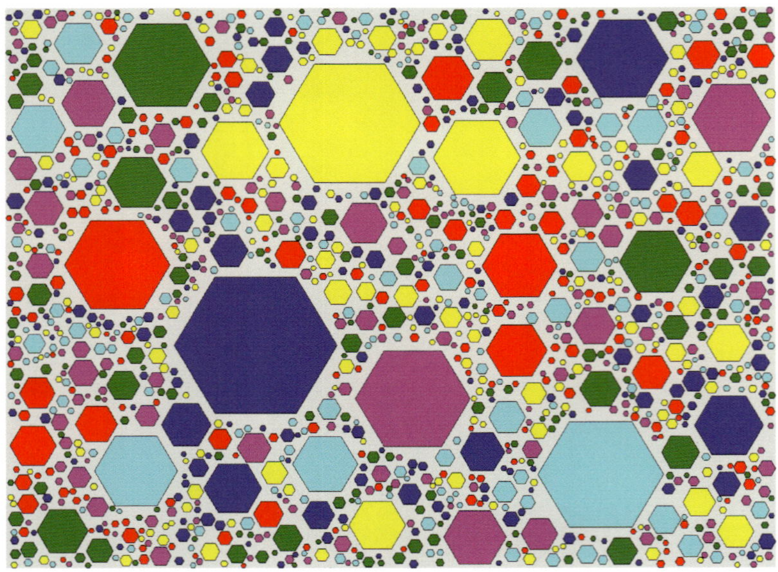

Fig. 9.20   ZNIOUAVUXZPGM $(PF = 0.600, CP = 0.812)$.

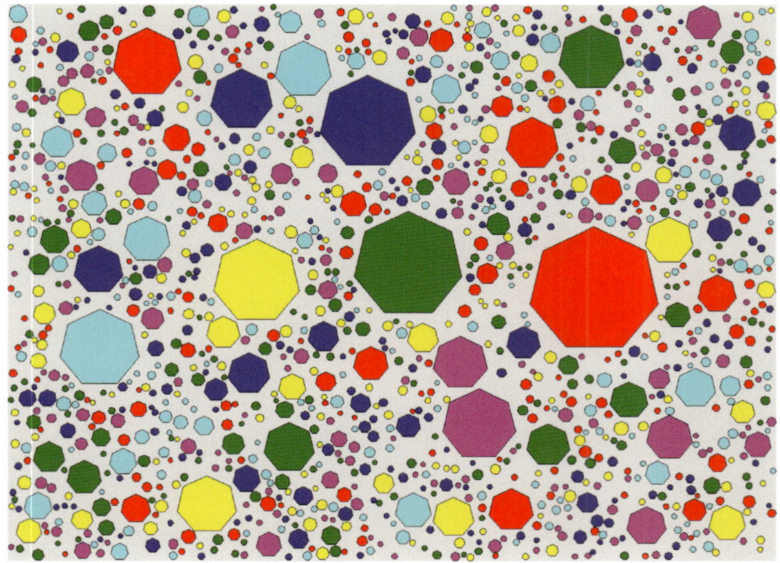

Fig. 9.21   ZOUUEIUNDZFAM ($PF = 0.439, CP = 0.833$).

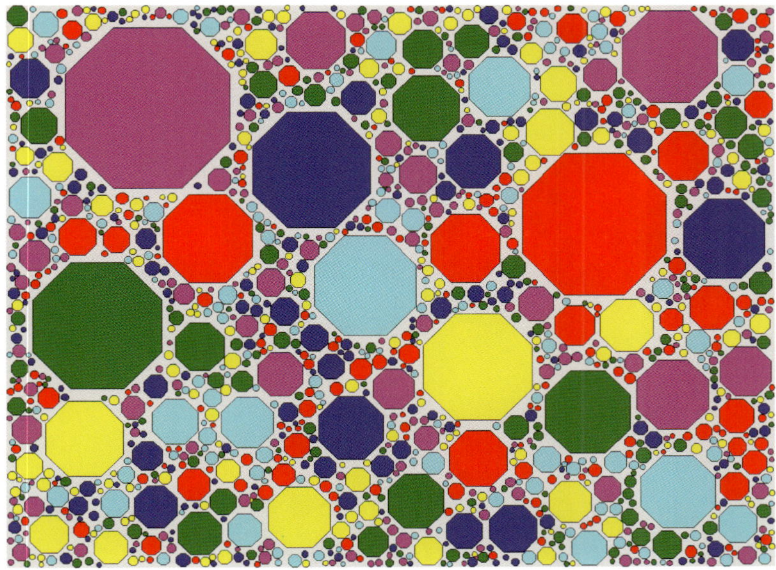

Fig. 9.22   ZPYKFWNDKZVKM ($PF = 0.739, CP = 0.817$).

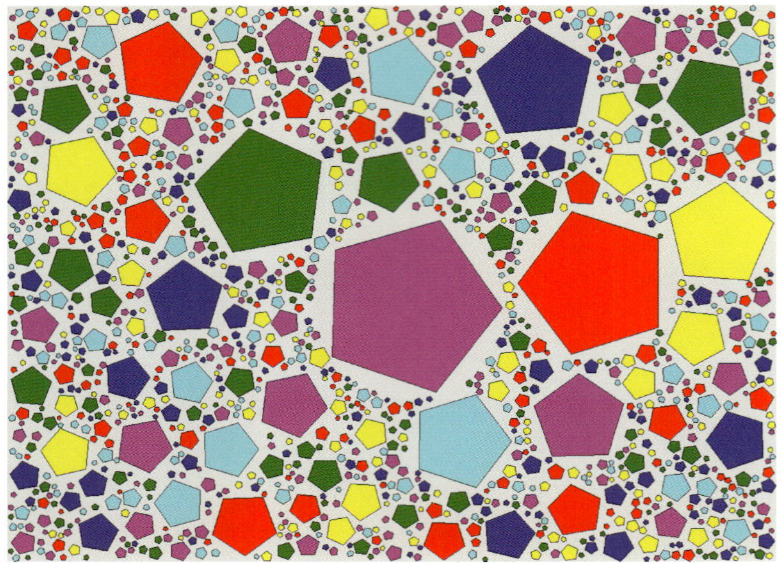

Fig. 9.23   ZQODYFPKWZGGM $(PF = 0.564, CP = 0.824)$.

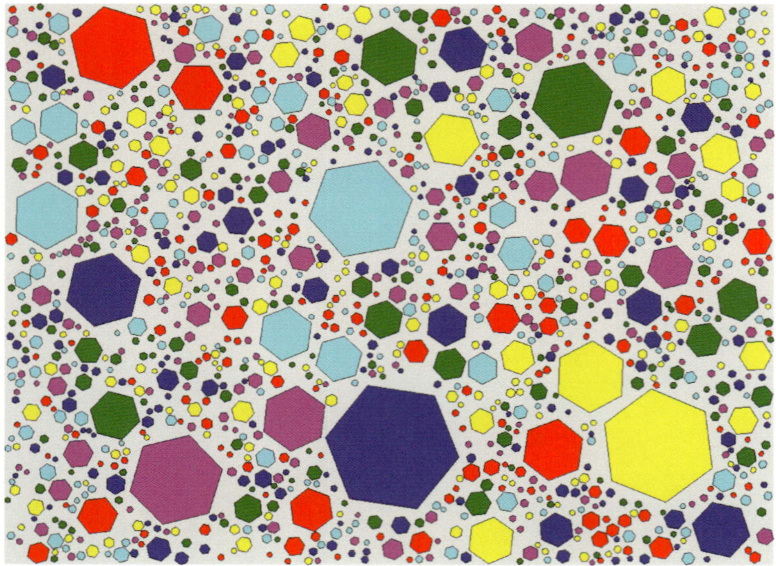

Fig. 9.24   ZRFXVGAJWZMCM $(PF = 0.479, CP = 0.826)$.

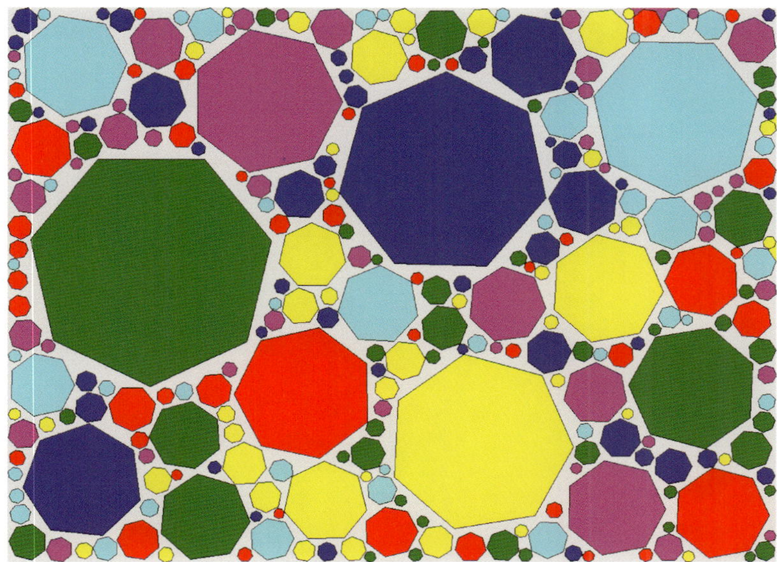

Fig. 9.25 **ZSTVBYSGKZYYM** $(PF = 0.790, CP = 0.906)$.

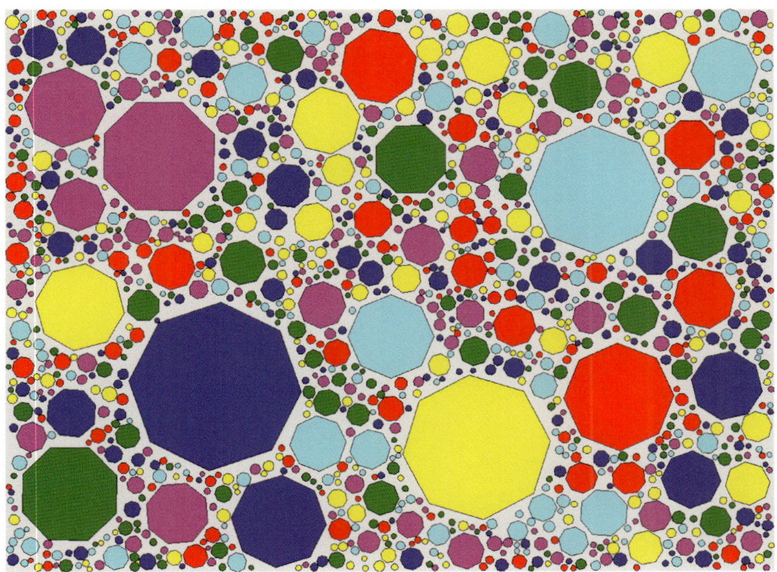

Fig. 9.26 **ZTKZNVMVEZOHM** $(PF = 0.679, CP = 0.812)$.

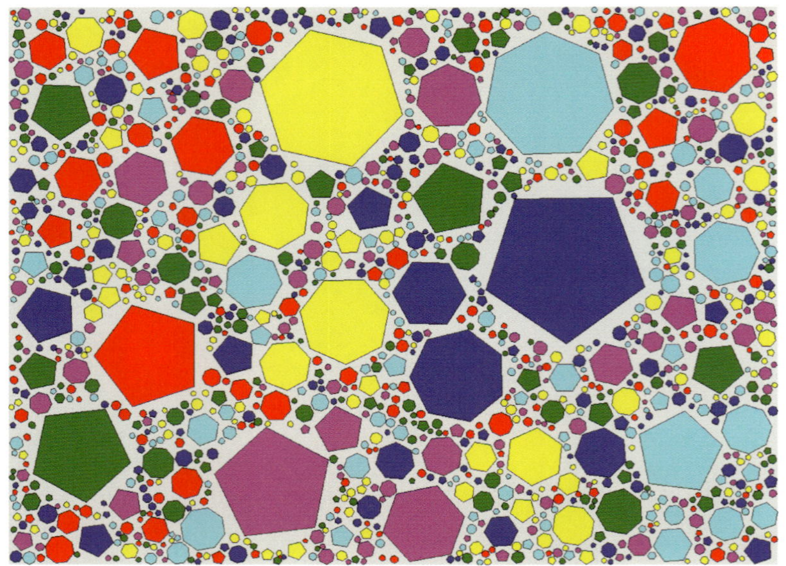

Fig. 9.27   **ZUJYEEHUGZUIM** $(PF = 0.639, CP = 0.813)$.

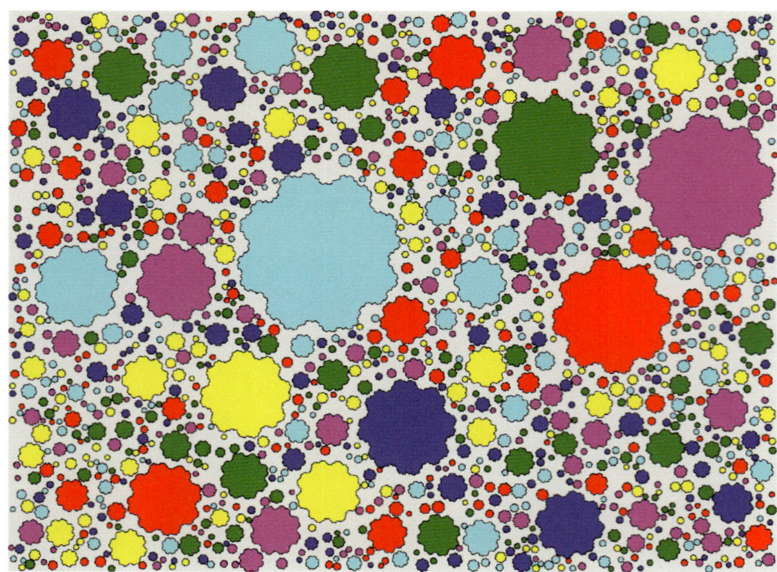

Fig. 9.28   **ZVNFZJKWYZEDM** $(PF = 0.593, CP = 0.766)$.

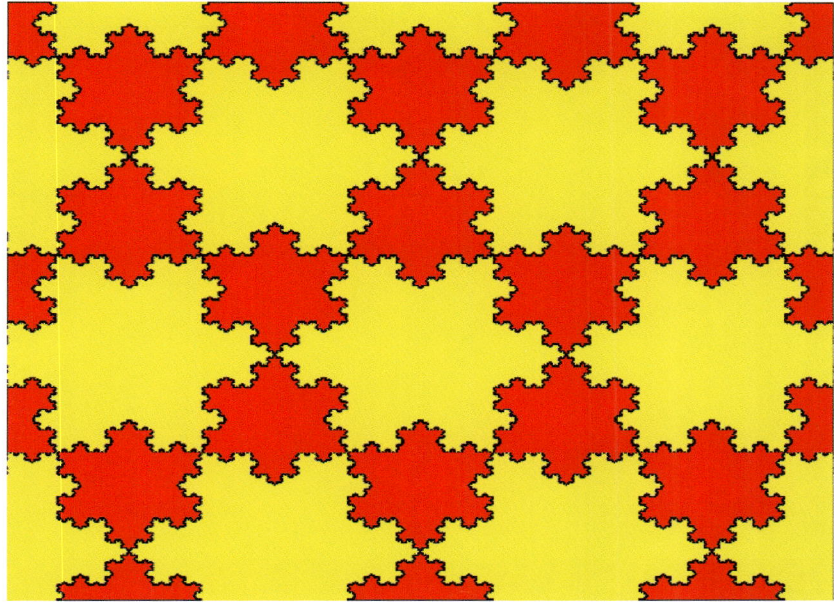

Fig. 9.29 A regular tessellation of the plane using two sizes of hexagonal Koch snowflakes.

each of which has a length reduced by a factor of $4 + 2\sqrt{2} \approx 6.828$, and repeating the process five times to give a fractal curve with a dimension of $D = \frac{\log(8)}{\log(4+2\sqrt{2})} \approx 1.0824$. Thus this figure is a kind of 'metafractal' or a 'fractal within a fractal' since the individual tiles are fractals, while there is a fractal distribution of tile sizes. Not only that, but the gray gasket has a fractal dimension given by $D = 2/c$ or $D \approx 1.85$ for Fig. 9.28.

It is also possible to tessellate the plane with two sizes of hexagonal Koch snowflakes as shown in Fig. 9.29. Do you think this regular fractal tiling is more or less elegant than the random tiling in Fig. 9.28? Perhaps the regular tiling would be more interesting if the tiles had a wider variety of colors. Such a tiling could be made into a metafractal by replacing each Koch snowflake with seven smaller snowflakes each of which in turn is replaced by seven even smaller ones and so forth to the limit of the graphics resolution [Burns (1994)].

A final example of a Shier tiling is shown in Fig. 9.30. It is difficult to avoid the illusion that you are viewing a large number of cubic blocks of

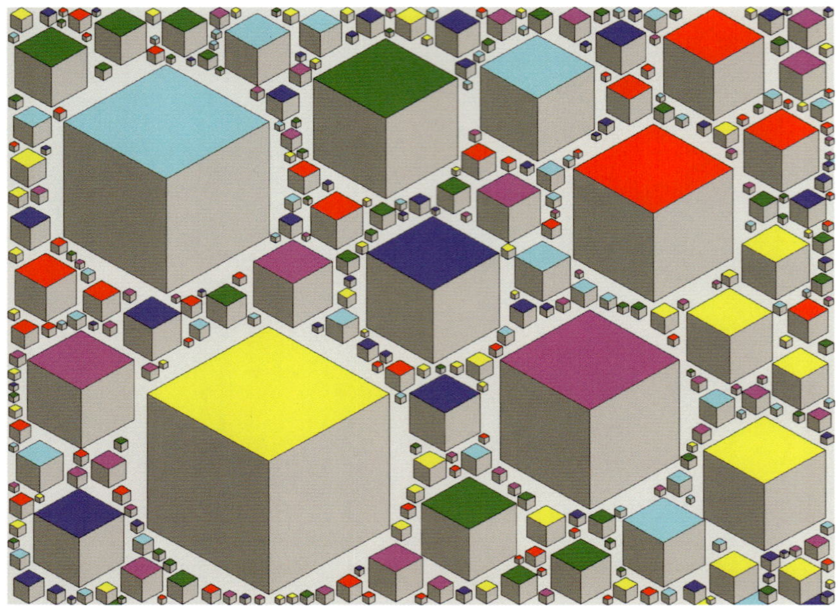

Fig. 9.30   `ZWCSNPBCYZVZM` $(PF = 0.773, CP = 0.854)$.

different sizes scattered across a landscape. However, if you look closely, you will see that each cube is actually just a hexagon with black lines connecting three of the vertices to its center and with the three resulting rhombuses colored to make them appear like blocks with colored tops illuminated by light from the right. In fact, it is similar to the case in Fig. 9.20 but with a different coloring scheme. It adds to the illusion that the larger hexagons are toward the bottom and left making them appear closer.

Figure 9.31 shows another case of fractal cubes but in a larger size and higher resolution. The background gasket color has been changed from gray to green to give the illusion of a cityscape with abundant grass viewed from above. Perhaps it reminds you of tombstones of various sizes in a well manicured but somewhat disordered cemetery.

We have now moved firmly from the realm of science and mathematics into the realm of art, and so perhaps this is a good place to stop and leave it for you to develop the ideas more fully and to produce your own examples of elegant fractal patterns.

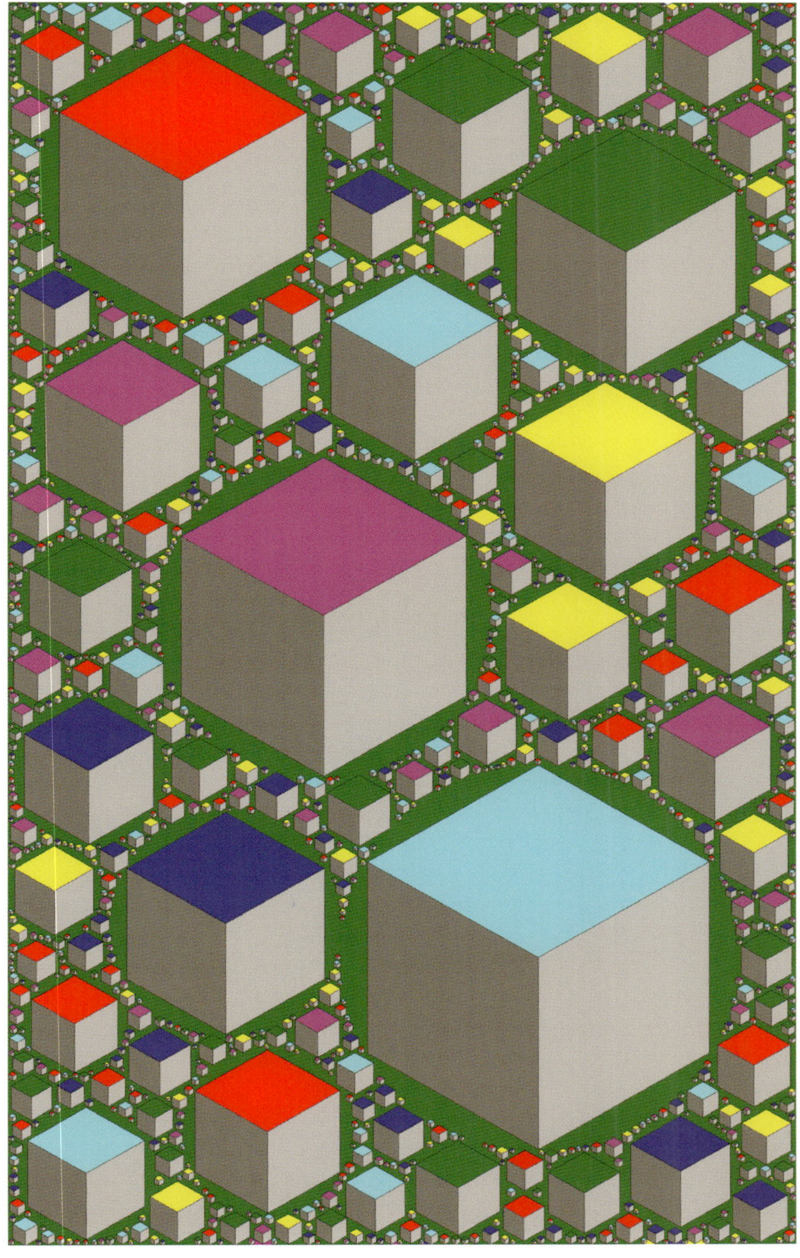

Fig. 9.31 ZWONPIOVGZQWM ($PF = 0.793, CP = 0.920$).

Chapter 10

Epilog

This short chapter will not introduce any new fractal types, but rather it will describe the computer program used to produce the images in the book. It will also summarize the state of the art in automating the selection of elegant fractal images. Finally, it will offer suggestions for future work and possible additional books on other forms of elegance.

10.1 Computer Software

Nearly all the figures in this book were produced by a single computer program called *Elegant Fractals* and written using the *PowerBASIC Console Compiler*. Non-programmers should be able to run the compiled program on any modern *Microsoft Windows* personal computer and reproduce any of the figures in the book whose caption contains a code or produce an unlimited number of other similar images that are different and never before seen. Since the program was written in parallel with the writing of the book, it grew by patching the program to accommodate an increasing diversity of fractal types, and thus some of the logic may be difficult for even an experienced programmer to follow.

Both the *PowerBASIC* source code (`elegfrac.bas`) and the compiled *Microsoft Windows* executable code (`elegfrac.exe`) are available from `http://sprott.physics.wisc.edu/fractals/elegant/`. This is a copyrighted program, but you are free to use it for noncommercial purposes and to modify it for your own use. It is distributed as-is without warranty of any kind and without technical support. You will also find at the above URL versions of all the figures in the book in **png** format. The program will allow you to make higher-resolution versions of any of the figures by just typing their code into the program and specifying the desired width and height of the plot in pixels.

The program was developed using *Microsoft Windows 10* and will be updated as necessary to ensure compatibility with future *Microsoft Windows* versions at least for the foreseeable future. There are no current plans to port the program to other languages or operating systems, but you may do that if you wish. Each time the program is compiled, it is given a new version number, which is just the compilation date, and so you can check that you have the most recent version. There is no tracking or documentation of the changes that were made or will be made between the many successive versions of the program.

The program has not been optimized for speed, and it often examines thousands of cases before finding one that is deemed sufficiently elegant to warrant saving it as a graphic image. Thus the usual search procedure involves running the program overnight and then examining the several dozen **bmp** graphics files that are typically produced. Their filenames are the codes that allow you to identify the parameters that were used and to replicate the images at higher resolution if desired. A list of the codes for all the elegant cases found in a search along with the date and time and values of IE, PF, CP, and SK are saved in the file **elegfrac.txt**. Some of the saved quantities are omitted or set to zero for certain fractal types where their value is meaningless or not useful.

When the program is launched, it opens a text window as shown in Fig. 10.1. There is no mouse or touch-screen support, and all the commands are entered from the keyboard. The opening screen responds to five keyboard commands:

- H(elp) opens a text box with an abbreviated version of these instructions.
- M(odify) opens a screen that allows you to change many of the default values used by the program. Probably the most useful of these values are the width and height of the plot in pixels (default 1200×900). You can retain the default values of the parameters by pressing **ENTER**.
- S(earch) allows you to search for new fractal images. This will open a screen as shown in Fig. 10.2 that asks the type of system to search with 26 options (**A** to **Z**). Pressing = will allow you to search for a range of types with a chosen starting and ending letter, or press **ENTER** to cause the program to search all types.
- V(erify) opens a screen that asks you to type the 13- or 14-character code for an image from the book or one you have found in a previous

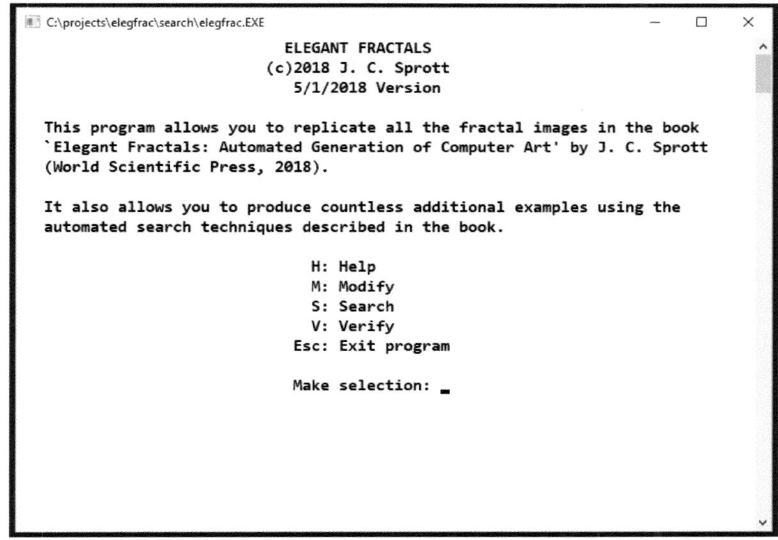

Fig. 10.1 Opening screen for the *Elegant Fractals* program.

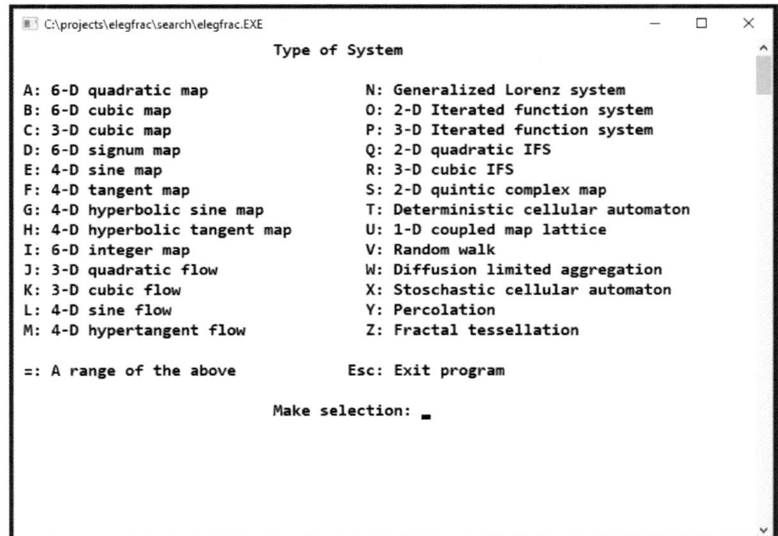

Fig. 10.2 Screen for the *Elegant Fractals* program that allows you to select the fractal type to search.

search. This allows you to replicate the image at higher (or lower) resolution or to calculate quantities such as the fractal dimension and Lyapunov exponent for some of the fractal types.

- Esc(ape) closes the program and exits.

10.2 Additional Examples

If you are impatient or do not want to tie up your computer searching for new elegant fractals, Table 10.1 gives the codes for 104 additional examples (four of each type) for you to explore. They have been selected for their elegance and could have served as additional figures in the book. You may want to create and save your own list of favorites or trade them with your

Table 10.1   Codes for additional examples of elegant fractals.

ACRDXJIEEGJEO	AGDGKHKPHLLOQ	AHMGWPQSWLWMV	ANVMXNMSUFUHZ
BMEGPNWKGUEOU	BNXPAMLOVFTGS	BSCMQQZIEOYFO	BWHBUXGTELMFT
CEMGZRYFKITXK	CIOWACVTJCXEU	CIZBFDFVOYHCH	CYKDNQCWRAVNK
DBLHNOMNXQFOG	DPSOMAHQHNLLB	DQMFZTFHQMRNI	DTLJNSYCIPWJI
EDZMDOCEAFDWB	ELHUJUAJTWEFS	EPTMOIBNFCFRI	EQTBGITOUTEJP
FDZNREQYJXFAI	FGFDEQITMNJCM	FLNMKFXYDXHUH	FUXAESCAHGPDW
GFJSJMHBIDNMC	GGRTVEMYSFDKM	GQPIYPCSEKIKN	GXSEXEQKOWBLF
HADOGDTNAASRG	HDZXANPFTRCKG	HJJXRAVTGHTQC	HPWXDGTRWJCEL
IDQLQOKHUKSMP	IGSFPKPGULNOP	IKIFPJSPOHTOJ	ITIXLJLGRJQRI
JETGWXODKJAUJ	JGSWGVSNNPJOY	JLJXHMIMOJYAI	JRALANUMNXJGE
KFLGWADSISNUA	KLLMNODREUPGF	KPFCKKIAQJNMC	KRELTFJFTHYBE
LEYWJBNECSDPG	LLBBIMFGATDTZ	LLETEBWEJUARO	LLWGKKGCRFJJS
MEYXISFGGRDBQ	MFXVCTSHVJEUB	MIFBISNJFTAIF	MLTJMKOAKZGCA
NMGCYPROTPEUR	NNRNEETSHUIPJ	NSJASFUAZTFEX	NTPWJGUREQBZV
OBLHKQLFTMUMH	ODLPJHLGENIOG	OMPPGRIQNPMIQ	OZRJXKMEFLLME
PLKPQHMSNJCPR	PMRQMMMGLEKKM	PNQGMCONPRQML	PYQJPNSKLXJMQ
QHMDKQGIHRGWM	QHZKZNRKTEOQH	QJPFNINLIXGZX	QNLGSLAPSOIOE
RFAFIIRFCNSKW	RKQXXTHFJQJUJ	RPMCCRNCCQMMS	RTKGDTKNCQSAJ
SIRMMEWMMWUMM	SJSMMSAMMIMMM	SMMBIMMJWMMTW	SMMYQMMIJMMLM
TAGFZADOVWVZM	TASXVZFVLSGAM	TBODKLEZQEVRM	TCVZWJTITSZTM
UBWYKUIUWVDMY	UBYQIXCQSUYLV	UCUAQCRVABRZC	UZUZRWTXJUFGZ
VAHTKHDGMMMMM	VAMAMSGVMMMMM	VCISAUYGMMMMM	VCQSYWYMMMMMM
WABABGNNLKHRM	WAWWQTYQKBIQC	WCAVQWLLCMDOW	WCIAUKBQUZUEM
XAINIZHKGKFON	XEQEDDXAEVSJO	XOYLGDHSKNPAR	XXWKEKFUHEDQW
YMIWDAPRQTXHI	YMMWHJQJDLLMH	YMSRVFFVNKQFK	YMXICTNHZYVGK
ZFRCNWAPKYMMM	ZHYYOCKZEFFMM	ZJRNNIFVLHFDM	ZWHUNOJXYZYIM

friends. Can you find ones that are more elegant than those shown in this book?

10.3 Extensions of the Method

The *infinite monkey theorem* states that a monkey hitting keys at random on a typewriter (or computer keyboard) for an infinite amount of time will almost surely reproduce the complete works of William Shakespeare (or any other piece of literature) [Isaac (1995)]. Of course the time required greatly exceeds the age of the Universe. On the other hand, a monkey with a paintbrush has a realistic of chance of producing an image that someone would consider 'artistic.' Thus it is not surprising that a modern personal computer with a minimum of human intervention can produce elegant fractal art [Sprott (2000, 2001, 2004)].

Elegance is one of those concepts, which, like beauty, is in the eye of the beholder and is difficult to define. Nevertheless, there is general agreement at least about extreme cases of elegance and inelegance. It is remarkable that certain easily calculated quantities for the various images correlate strongly with one's subjective judgement of their elegance. That raises the possibility of programming the computer to make aesthetic judgements that at least offer assistance to the human artist or mathematician in search of interesting new examples of fractals.

This book has provided a mere glimpse of the possibilities. Quantities such as fractal dimension, Lyapunov exponent, pixel fractal, cluster probability, and skewness have all proved useful in weeding out the vast majority of fractals that most people would not consider elegant. And the ideas have been applied to a tiny sliver of the countless mathematical models that produce fractals. There is ample room for you to explore other models and to invent and test other quantities that are equally or even more useful in assessing their elegance.

The problem is also amenable to modern machine learning techniques. Imagine training a machine on the best artworks of the World and then having it produce millions of candidate images and retain only those few that are likely to have wide human appeal. Perhaps computers will develop their own artistic style and preferences that are different from those of humans. One cannot escape the idea that we may be on the cusp of a new art form dominated by the machines that we have created. Whether this is something to fear or something to embrace is yet an open question for you to ponder.

10.4 Other Elegant Concepts

This book is a sequel to the earlier book *Elegant Chaos: Algebraically Simple Chaotic Flows* [Sprott (2010)]. Certainly an elegant equation and an elegant fractal are different concepts and need to be defined and evaluated differently, but any physicist or mathematician can identify an elegant equation or an elegant theorem just as easily as an artist or art critic can identify an elegant painting or a poet can identify an elegant poem. Fractals bridge the fields of science, mathematics, and art perhaps more than any other topic.

Elegance occurs in countless other forms across the arts and sciences. Music and literature can be more or less elegant. You would probably rank the writings of William Shakespeare as more elegant than the words you have been reading here. Computer programs, mathematical theorems and their proofs, and electronic circuits can be elegant. In fact, the next book in this sequence, *Elegant Circuits: Simple Chaotic Oscillators* [Sprott and Thio (2020)] is already underway. An elegant circuit is one that performs a given function, in this case producing a chaotic signal with a bare minimum of simple components, just as one type of elegant equation has a minimum of simple terms and a chaotic solution.

A currently popular subject in nonlinear dynamics is *complexity* [Waldrop (1992)], and there are many elegant mathematical models that exhibit complex behavior including self-organization and pattern formation [Mitchell (2009)]. Thus a logical next book closely related to the first three might be entitled *Elegant Complexity: Simple Models with Complex Behavior*. Perhaps you are an expert in this field and would consider writing such a book or you know someone who would.

Beyond that, the possibilities are endless. Every field of study and every form of art has elegant models, calculations, and examples that could be consolidated into a series of books embracing ever widening domains of scientific and artistic endeavours. We should learn from one another and grow to appreciate the elegance that others see in domains that are far removed from our own and share with others the elegance that occurs in our own fields of interest and expertise, which can only lead to a more connected and tolerant World.

Bibliography

Aks, D. J. and Sprott, J. C. (1996). Quantifying aesthetic preference for chaotic patterns, *Emp. Stud. Arts*, pp. 1–16.

Albers, D. J. and Alexanderson, G. L. (2008). Benoit Mandelbrot: In his own words, *Mathematical People: Profiles and Interviews* (A. K. Peters, Wellesley, MA), pp. 214.

Algaba, A., Fernández-Sánchez, G., Merino, M., and Rodríguez-Luis, J. J. (2013). Chen's attractor exists if Lorenz repulsor exists: The Chen system is a special case of the Lorenz system, *Chaos* **23**, pp. 033108-1–6.

Appel, K. and Haken, W. (1989). *Every Planar Map is Four Colorable: Contemporary Mathematics* **98** (American Mathematical Society, Providence, RI).

Arfken, G. (1985). *Mathematical Methods for Physicists*, 2nd edition (Academic Press, Orlando, FL).

Armstrong, M. A. (1988). *Groups and Symmetry* (Springer-Verlag, New York).

Atkinson, K., Han, W., and Stewart, D. E. (2009). *Numerical Solution of Ordinary Differential Equations* (Wiley, New York).

Ball, R., Nauenberg, M., and Witten, T. A. (1984). Diffusion-controlled aggregation in the continuum approximation, *Phys. Rev. A* **29**, pp. 2017–2020.

Barnsley, M. (1988). *Fractals Everywhere* (Academic Press, Boston).

Barnsley, M. F. and Hurd, L. P. (1993). *Fractal Image Compression* (A. K. Peters, Wellesley, MA).

Branner, B. (1988). The Mandelbrot set, In *Chaos and Fractals: The Mathematics behind the Computer Graphics*, eds. R. Devaney and L. Keen (American Mathematicl Society, Providence, RI), pp. 75–105.

Bertin, G. and Lin, C. C. (1996). *Spiral Structure in Galaxies: A Density Wave Theory* (MIT Press, Cambridge, MA).

Broadbent, S. and Hammersley, J. (1957). Percolation processes: 1. Crystals and mazes, *P. Camb. Philol. Soc.* **53**, pp. 629–641.

Brummitt, C. D. and Sprott, J. C. (2009). A search for the simplest chaotic partial differential equation, *Phys. Lett. A* **373**, pp. 2717–2721.

Burns, A. (1994). Fractal tilings, *Math. Gaz.* **72**, pp. 193–196.

Cajori, F. (1993). *A History of Mathematical Notations* (Dover, New York).

Cattaneo, G., Finellim M., and Margara, L. (2000). Investigating topological chaos by elementary cellular automata dynamics, *Theor. Comput. Sci.* **244**, pp. 219–241.

Chapman, R. S. and Sprott, J. C. (2005). *Images of a Complex World: The Art and Poetry of Chaos* (World Scientific, Singapore).

Chen, G. and Ueta, T. (1999). Yet another chaotic attractor, *Int. J. Bifurcat. Chaos Appl. Sci. Eng.* **9**, pp. 1465–1466.

Chlouverakis, K. E. and Sprott, J. C. (2005). Chaotic hyperjerk systems, *Chaos, Solitons & Fractals* **28**, pp. 739–746.

Clifford, P. and Sudbury, A. (1973). A model for spatial conflict, *Biometrika* **60**, pp. 581–588.

Cortie, M. (1992). The form, function, and synthesis of the molluscan shell, In *Spiral Symmetry*, eds. Hargittai, I. and Pickover, C. A. (World Scientific, Singapore), pp. 356–357.

Coxeter, H. S. M. (1968). The problem of Apollonius, *Am. Math. Mon.* **75**, pp. 5–15.

Coxeter, H. S. M. (1973). *Regular Polytopes*, 3rd edition (Dover, New York).

Dodds, P. S. and Weitz, J. S. (2002). Packing-limited growth, *Phys. Rev E* **65**, pp. 056108-1–6.

Efros, A. L. and Shklovskii, B. I. (1976). Critical behavior of conductivity and dielectric constant near the metal-non-metal transition threshold, *Phys. Status Solidi B* **76**, pp. 475–485.

Eichhorn, R., Linz, S. J., and Hänggi, P. (1998). Transformations of nonlinear dynamical systems to jerky motion and its application to minimal chaotic flows, *Phys. Rev. E* **58**, pp. 7151–7164.

Eichhorn, R., Linz, S. J., and Hänggi, P. (2002). Simple polynomial classes of chaotic jerky dynamics, *Chaos, Solitons & Fractals* **13**, pp. 1–15.

Ennis, C. (2016). (Always) room for one more, *Math Horizons* **23** (3), pp. 8–12.

Escher, M. C. (1958). *Regelmatige Vlakverdeling (Regular Division of the Plane)* (De Roos Foundation, Utrecht).

Fernández-Gracia, J., Suchecki, K., Ramasco, J. J., Miguel, M. S., and Eguiluz, V. M. (2014). Is the voter model a model of voters?, *Phys. Rev. Lett.* **112**, pp. 158701-1–5.

Field, M. and Golubitsky, M. (1992). *Symmetry in Chaos* (Oxford University Press, New York).

Freeman, H. (1974). Computer processing of line-drawing images, *ACM Comput. Surv.* **6**, pp. 57–97.

Gardner, M. (1970). The fantastic combinations of John Conway's new solitaire game "life," *Sci. Am.* **223** (4), pp. 120–123.

Gardner, R. H., Milne, B. T., Turner, M. G., and O'Neill, R. V. (1987). Natural models for the analysis of broad-scale landscape pattern, *Landscape Ecol.* **1**, pp. 5–18.

Gear, C. W. (1971). *Numerical Initial Value Problems in Ordinary Differential Equations* (Prentice-Hall, Englewood Cliffs, NJ).

Geist, K., Parlitz, U., and Lauterborn, W. (1990). Comparison of different methods for computing Lyapunov exponents, *Prog. Theor. Phys.* **83**, pp. 875–893.

Gilbert, E. N. (1967). Random plane networks and needle-shaped crystals, In *Applications of Undergraduate Mathematics in Engineering*, ed. Noble, B. (Macmillan, New York).

Gilmore, R. and Letellier, C. (2007). *The Symmetry of Chaos* (Oxford University Press, New York).

Goldberg, D. (1991). What every computer scientist should know about floating-point arithmetic, *ACM Comput. Surv.* **23**, pp. 5–48.

Grassberger, P. and Procaccia, I. (1983). Measuring the strangeness of strange attractors, *Phys. Nonlinear Phenom.* **9**, pp. 189–208.

Gray, A. (1901). *Treatise on Physics, Volume 1* (Churchill, London).

Gray, N. H., Anderson, J. B., Deving, J. D., and Kwasnik, J. M. (1976). Topological properties of random crack networks, *Math. Geol.* **8**, pp. 617–626.

Griffiths, D. F. and Higham, D. J. (2010). *Numerical Methods for Ordinary Differential Equations* (Springer-Verlag, London).

Grünbaum, B. and Shephard, G. C. (1987). *Tilings and Patterns* (Freeman, New York).

Hammersley, J. M. (1983). Origins of percolation theory, *Ann. Isr. Phy.* **5**, pp. 47–57.

Harte, D. (2001). *Multifractals* (Chapman and Hall, London).

Herega, A. (2015). Some applications of the percolation theory: Brief review of the century beginning, *J. Mater. Sci. Eng. A* **5**, pp. 409–414.

Hénon, M. (1976). A two-dimensional mapping with a strange attractor, *Comm. Math. Phys.* **50**, pp. 69–77.

Hirsch, M. W., Smale, S., and Devaney, R. L. (2004). *Dynamical Systems, and an Introduction to Chaos*, 2nd edition (Elsevier/Academic Press, Amsterdam).

Hoover, W. G., Sprott, J. C., and Patra, P. K. (2015). Ergodic time-reversible chaos for Gibbs' cannonical oscillator, *Phys. Lett. A* **379**, pp. 2935–2940.

Hornik, K., Stinchocombe, M., and White, H. (1989). Multilayer feedforward networks are universal approximators, *Neural Networks* **2**, pp. 359–366.

Hutchinson, J. (1981). Fractals and self similarity, *Indiana U. Math. J.* **30**, pp. 713–747.

Isaac, R. E. (1995). *The Pleasures of Probability* (Springer, New York).

Ivic, A. (1985). *The Riemann Zeta Function* (John Wiley & Sons, New York).

Julia, G. (1918). Memoire sur l'itération des fonctions rationnelles, *Journal Mathématiques Pur et Appliquées* **8**, pp. 47–245.

Kaneko, K. (1984). Pereiod-doubling of kink-antikink patterns, quasi-periodicity in antiferro-like structures and spatial intermittency in coupled map lattices – toward a prelude to a "field theory of chaos," *Prog. Theor. Phys.* **72**, pp. 480–486.

Kaplan, J. and Yorke, J. (1979). Chaotic behavior of multidimensional difference equations, In *Functional Differential Equations and Approximation of Fixed Points, Lecture Notes in Mathematics*, Vol. 730, eds. H.-O. Peitgen and H.-O. Walther, pp. 228–237 (Springer, Berlin).

Karatsuba, A. A. and Voronin, S. M. (1992). *The Riemann Zeta-Function* (de Gruyter, Berlin).

Kesten, H. (1987). *Percolation Theory and Ergodic Theory of Infinite Particle Systems* (Springer, New York).

Klee, P. (1998–2004). *Catalogue Raisonné* (9 vols.) Edited by the Paul Klee Foundation, Museum of Fine Arts, Berne (Thames and Hudson, New York).

Knuth, D. E. (1997). *Seminumerical Algorithms*, 3rd edition, Vol. 2 of *The Art of Computer Programming* (Addison-Wesley Longman, Reading, MA).

Knuth, D. E. (1998). *Sorting and Searching*, 3rd edition, Vol. 3 of *The Art of Computer Programming* (Addison-Wesley Longman, Reading, MA).

Kuramoto, Y. and Tsuzuki, T. (1976). Persistent propagation of concentration waves in dissipative media far from thermal equilibrium, *Prog. Theor. Phys.* **55**, pp. 356–369.

Kuznetsov, S. P. and Pikovsky, A. S. (1985). Universality of period doubling bifurcations in one-dimensional dissipative media, *Izvestija VUS, Radiofizika* **28**, pp. 308–320.

L'Ecuyer, P. (1988). Efficient and portable combined random number generators, *Commun. ACM* **31**, pp. 752–774.

Liggett, T. M. (1985). *Intereacting Particle Systems* (Springer-Verlag, New York).

Liggett, T. M. (1999). *Stochastic Interacting Systems: Contact, Voter and Exclusion Processes* (Springer-Verlag, New York).

Livio, M. (2002). *The Golden Ratio: The Story of PHI, the World's Most Astonishing Number* (Random House, New York).

Lodi, A., Martello, S., and Monaci, M. (2002). Two-dimensional packing problems: A survey, *Eur. J. Oper. Res.* **141**, pp. 241–252.

Lorenz, E. N. (1963). Deterministic nonperiodic flow, *J. Atmos. Sci.* **20**, pp. 130–141.

Lorenz, E. N. (1993). *The Essence of Chaos* (University of Washington Press, Seattle).

Lozi, R. (1978). Un attracteur étrange? Du type attracteur de Hénon, *J. Phys-Paris)* **39**, pp. 9–10.

Mackisack, M. S. and Miles, R. E. (1996). Homogeneous rectangular tessellations, *Adv. Appl. Probab.* **28**, pp. 993–1013.

Malasoma, J. -M. (2000). What is the simplest dissipative chaotic jerk equation which is parity invariant?, *Phys. Lett. A* **264**, pp. 383–389.

Mandelbrot, B. B. (1982). *The Fractal Geometry of Nature* (Freeman, New York).

Mauldin, R. D. and Urbanski, M. (1998). Dimension and measures for a curvilinear Sierpiński gasket of Apollonian packings, *Adv. Math.* **136**, pp. 26–38.

Menger, K. (1928). *Dimensionstheorie* (B. G. Teubner Publishers, Leipzig).

Milne, W. E. (1970). *Numerical Solution of Differential Equations* (Dover, New York).

Mitchell, M. (2009). *Complexity: A Guided Tour* (Oxford, New York).

Moore, E. F. (1962). Machine models of self reproduction, *American Mathematical Society Proceedings of Symposia in Applied Mathematics* **14**, pp. 17–33.

Munmuangsaen, B. and Srisuchinwong, B. (2009). A new five-term simple chaotic attractor, *Phys. Lett. A* **373**, pp. 4038–4043.

Nazzi, F. (2016). The hexagonal shape of the honeycomb cells depends on the construction behavior of bees, *Sci. Rep.* **6**, pp. 28341-1–6.

Park, S. K. and Miller, K. M. (1988). Random number generators: Good ones are hard to find, *Commun. ACM* **31**, pp. 1192–1201.

Peitgen, H.-O. and Richter, P. H. (1986). *The Beauty of Fractals: Images of Complex Dynamical Systems* (Springer, Berlin).

Peitgen, H.-O., Jürgens, H., and Saupe, D. (2004). *Chaos and Fractals: New Frontiers of Science*, 2nd edition (Springer-Verlag, New York).

Penrose, R. (1974). The role of aesthetics in pure and applied mathematical research, *B. I. Math. Appl.* **10**, pp. 266–271.

Pesin, Ya. B. (1977). Characteristic Lyapunov exponents and smooth ergodic theory, *Russ. Math. Surv.* **32**, pp. 55–114.

Press, W. H. and Teukolsky, S. A. (1992). Portable random number generators, *Comput. Phys.* **6**, pp. 522–524.

Press, W. H., Teukolsky, S. A., Vetterling, W. T., and Flannery, B. P. (2007). *Numerical Recipes: The Art of Scientific Computing*, 3rd edition (Cambridge University Press, Cambridge).

Rényi, A. (1970). *Probability Theory* (North Holland, Amsterdam).

Richards, R. (1999). The subtle attraction: Beauty as a force in awareness, creativity, and survival, In *Affect, Creative Experience, and Psychological Adjustment*, ed. S. W. Russ, pp. 195–219 (Brunner/Mazel, Philadelphia).

Rössler, O. E. (1976). An equation for continuous chaos, *Phys. Lett. A* **71**, pp. 155–157.

Rössler, O. E. (1979). Continuous chaos – four prototype equations, *Ann. New York Acad. Sci.* **316**, pp. 376–392.

Saberi, A. A. (2015). Recent advances in percolation theory and its applications, *Phys. Rep.* **578**, pp. 1–32.

Sahimi, M. (1994). *Applications of Percolation Theory* (Taylor and Francis, London).

Schmidhuber, J. (2015). Deep learning in neural networks: An overview, *Neural Networks* **61**, pp. 85–117.

Schot, S. H. (1978). Jerk: The time rate of change of acceleration, *Am. J. Phys.* **46**, pp. 1090–1094.

Shier, J. and Bourke, P. (2013). An algorithm for random fractal filling of space, *Comput. Graph. Forum* **32** (8), pp. 89–97.

Sierpiński, W. (1916). Sur une courbe cantorienne qui contient une image biunivoque et continue de toute courbe donnée, *Comptes Rendus Acad. Sci. Paris* **162**, pp. 629–632.

Sivashinsky, G. I. and Michelson, D. M. (1980). On the irregular wavy flow of a liquid film flowing down a vertical plane, *Prog. Theor. Phys.* **63**, pp. 2112–2117.

Sprott, J. C. (1992). Simple programs create 3-D images, *Comput. Phys.* **6**, pp. 132–138.

Sprott, J. C. (1993a). Automatic generation of strange attractors, *Comput. Graph.* **17**, pp. 325–332.

Sprott, J. C. (1993b). *Strange Attractors: Creating Patterns in Chaos* (M & T Books, New York).

Sprott, J. C. (1994a). Some simple chaotic flows, *Phys. Rev. E* **50**, pp. R647–650.

Sprott, J. C. (1994b). Automatic generation of iterated function systems, *Comput. Graph.* **18**, pp. 417–425.

Sprott, J. C. (1996). Strange attractor symmetric icons, *Comput. Graph.* **20**, pp. 325–332.

Sprott, J. C. (1997a). Simplest dissipative chaotic flow, *Phys. Lett. A* **228**, pp. 271–274.

Sprott, J. C. (1997b). Some simple chaotic jerk functions, *Am. J. Phys.* **65**, pp. 537–543.

Sprott, J. C. (1998). Artificial neural net attractors, *Comput. Graph.* **22**, pp. 143–149.

Sprott, J. C. (2000). Automatic generation of fractal art, *YLEM Newsletter* **20**, pp. 10–12.

Sprott, J. C. (2001). Can a computer produce and critique art?, *Leonardo* **34**, pp. 369.

Sprott, J. C. (2003). *Chaos and Time-Series Analysis* (Oxford University Press, Oxford).

Sprott, J. C. (2004). Can a monkey with a computer create art?, *Nonlin. Dynam. Psychol.* **8**, pp. 103–114.

Sprott, J. C. (2007). Maximally complex simple attractors, *Chaos* **17**, pp. 033124-1–6.

Sprott, J. C. (2010). *Elegant Chaos: Algebraically Simple Chaotic Flows* (World Scientific, Singapore).

Sprott, J. C. (2014). *Fractals Colouring Book* (Arcturus Publishing Limited, London).

Sprott, J. C. (2015). New chaotic regimes in the Lorenz and Chen systems, *Int. J. Bifurcat. Chaos*, **25**, pp. 1550033-1–7.

Sprott, J. C. and Pickover, C. A. (1995). Automatic generation of general quadratic map basins, *Comput. Graph.* **19**, pp. 309–313.

Sprott, J. C. and Rowlands, G. (2001). Improved correlation dimension calculation, *Int. J. Bifurcat. Chaos Appl. Sci. Eng.* **11**, pp. 1861–1880.

Sprott, J. C. and Thio, W. J. (2020). *Elegant Circuits: Simple Chaotic Oscillators* (World Scientific, Singapore).

Sprott, J. C. and Xiong, A. (2015). Classifying and quantifying basins of attraction, *Chaos* **25**, pp. 08301-1–7.

Sprott, J. C., Bolliger, J., and Mladenoff, D. J. (2002). Self-organized criticality in forest-landscape evolution, *Phys. Lett. A* **297**, pp. 267–271.

Stauffer, D. (1985). *Introduction to Percolation Theory* (Taylor and Francis, London).

Stewart, I. and Golubitsky, M. (1992). *Fearful Symmetry: Is God a Geometer?* (Blackwell, Cambridge, MA).

Titchmarsh, E. C. (1986). *The Theory of the Riemann Zeta-Function*, 2nd edition (Clarendon Press, Oxford).

Ulam, S. (1952). Random processes and transformations, *Proceedings of the International Congress on Mathematics* **2** (American Mathematical Society, Providence, RI), pp. 264–275.

Umberger, D. K. and Farmer, J. D. (1985). Fat fractals on the energy surface, *Phys. Rev. Lett.* **55**, pp. 661–664.

van der Schrier, G. and Maas, L. R. M. (2000). The diffusionless Lorenz equations; Shil'nikov bifurcations and reduction to an explicit map, *Phys. Nonlinear Phenom.* **141**, pp. 19–36.

Vepstas, L. (2007). An efficient algorithm for accelerating the convergence of oscillatory series, useful for computing the polylogarithm and Hurwitz zeta functions, *Numer. Algorithms* **47**, pp. 211–252.

von Koch, H. (1904). Sur une courbe continue sans tangente, obtenue par une construction géométrique élémentaire, *Ark. Mat.* **1**, pp. 681–704.

von Neumann, J. (1966). *Theory of Self-reproducing Automata*, ed. A. W. Burks (University of Illinois Press, Urbana, IL).

Waldrop, M. (1992). *Complexity: The Emerging Science at the Edge of Order and Chaos* (Simon and Schuster, New York).

Waller, I. and Kapral, R. (1984). Spatial and temporal structure in systems of coupled nonlinear oscillators, *Phys. Rev. A* **30**, pp. 2047–2055.

Watts, D. J. and Strogatz, S. H. (1998). Collective dynamics of 'small-world' networks, *Nature* **393**, pp. 440–442.

Witten, T. A. and Sander, L. M. (1983). Diffusion-limited aggregation, *Phys. Rev. B* **27**, pp. 5686–5697.

Wolf, A., Swift, J. B., Swinney, H. L., and Vastano, J. A. (1985). Determining Lyapunov exponents from a time series, *Phys. Nonlinear Phenom.* **16**, pp. 285–317.

Wolfram, S. (1983). Statistical mechanics of cellular automata, *Rev. Mod. Phys.* **55**, pp. 601–644.

Wolfram, S. (1986). Random sequence generation by cellular automata, *Adv. Appl. Math.* **7**, pp. 123–169.

Wolfram, S. (2002). *A New Kind of Science* (Wolfram Media, Champaign, IL).

Wolfram, S. (2017). *An Elementary Introduction to the Wolfram Language* (Wolfram Media, Champaign, IL).

Ziff, R. M. and Sapoval, B. (1986). The efficient determination of the percolation threshold by a frontier-generating walk in a gradient, *J. Phys. A-Math. Gen.* **19**, pp. L1169–1172.

Index

About the Author

JULIEN CLINTON SPROTT is Emeritus Professor of Physics at the University of Wisconsin–Madison. After a twenty-five year career in plasma physics, he became interested in chaos and fractals in 1988. He is the author of about five hundred technical papers and a dozen books including *Chaos and Time-Series Analysis* (Oxford, 2003) and *Elegant Chaos* (World Scientific, 2010). He has produced thirty-five hour-long videos of his popular public presentation of *The Wonders of Physics* and four commercial software packages. His award-winning Web site is at `http://sprott.physics.wisc.edu/`.

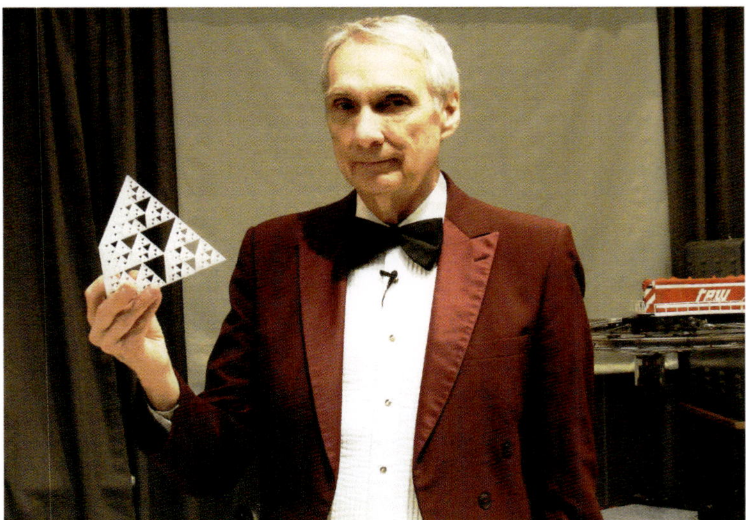

The author displays a 3-D printed Sierpiński tetrahedron during a public presentation of *The Wonders of Physics* in 2014.